PRAISE FOR **THE END OF KILLING**

"This book is a triumph. Rick Smith challenges what we think we know about everything: how we police our streets, how we protect our schools, how we fight our wars. And what he proves is that, in all of these fields, technology can help build a safer world with far less violence. For anyone in the business of trying to build a future that looks better than our present, this book contains ideas worth reading and absorbing."

CONGRESSMAN DAVID SCHWEIKERT

"If you can only kill, unless you're willing to kill everyone, then our power is defined by our foes' willingness to back down. Rick Smith challenges the notion that the only thing we ought to do is kill our opponents—and that idea has the potential to reshape how we think about violence, and ultimately, how we think about power itself. This is an important and thought-provoking book."

GENERAL STANLEY A. MCCHRYSTAL

"I've served in combat and lost friends overseas. In *The End of Killing*, Rick Smith asks the questions that hang in the minds of every war veteran: isn't there a better way? Do countless lives have to be lost in order to keep the peace? Couldn't we change how we fight wars using technology? This is a provocative book—but it provokes exactly the questions we must be asking if we want to build a more peaceful world."

CONGRESSMAN RUBEN GALLEGO

"Rick Smith is known as the 'Steve Jobs of policing technology'—and this book reflects his well-deserved reputation. Smith envisions a world free of gun violence. This is a must-read for those who share his bold vision to enhance safety in our communities!"

KATHY O'TOOLE, former chief of police, Seattle Police Department

"This is a brilliant, insightful, and provocative book. An absolutely essential argument for applying technology toward 'less lethal' solutions."

LT. COL. DAVE GROSSMAN, author of *On Killing* and *On Combat*

"The future belongs to those who build it. Rick Smith has spent his entire life building the future—it's because of him that police officers carry non-lethal weapons and it's because of those weapons that thousands of lives are saved each year. Now, Smith lets us into his mind and world, and he outlines a vision of the future that is safer, more just, and more peaceful. A stunning achievement!"

PETER H. DIAMANDIS, founder and executive chairman, XPRIZE Foundation; executive chairman, Singularity University

"As a former surgeon general, US Army Special Forces soldier, police officer, and trauma surgeon/EMS director, the consequences of bullets and lethal force have always weighed heavily on me and my colleagues. We have all aspired to see a non-lethal force option that could safely stop a threat and put an end to killing. Through persistence, passion, and the innovative use of technology, Rick Smith has accelerated and envisioned the future. Now he's given us a window, through his book, *The End of Killing*, to see a future we can create—and one that speaks to our highest aspirations."

RICHARD CARMONA, 17th Surgeon General of the United States

"*The End of Killing* should be required reading for everybody, but especially police practitioners, civil rights advocates, local policymakers, state legislators, and members of Congress. Rick Smith makes a powerful, aspirational argument for the use of exponentially advancing technology to dramatically reduce violence in our homes, schools, and communities. We now stand at a crossroads. Use technology to greatly reduce these tragic occurrences or continue to experience an absurd level of preventable violence. Smith lays out a logical path forward to reducing violence and many of its antecedents. It is up to the rest of us to take this disruptive idea—the end of killing—and make it a reality."

JIM BUEERMANN, former president of the National Police Foundation; founder, Future Policing Institute

THE
END
OF
KILLING

HOW OUR NEWEST
TECHNOLOGIES
CAN SOLVE HUMANITY'S
OLDEST PROBLEM

RICK SMITH

THE END OF KILLING

PAGE TWO
BOOKS

Cataloguing in publication information is available from Library and Archives Canada.
ISBN 978-1-989025-53-6 (hardcover)
ISBN 978-1-989025-54-3 (ebook)

Page Two
www.pagetwo.com

Edited by Amanda Lewis
Designed by Peter Cocking
Printed and bound in Canada by Friesens
Distributed in Canada by Raincoast Books
Distributed in the US and internationally
by Publishers Group West, a division of Ingram

19 20 21 22 23   5 4 3 2 1

Join the conversation on Twitter: #EndOfKilling
www.EndOfKilling.com

"If we can put a man on the moon and return him safely to Earth, why can't we put a man on the ground and take him safely to jail?"

CAPTAIN GREG MEYER, LAPD (RET.)

"Don't find fault, find a remedy."

HENRY FORD

This book is dedicated to the men and women in public safety and the military who risk their lives each day so the rest of us may sleep soundly each night.

CONTENTS

AUTHOR'S NOTE

TWENTY-FIVE YEARS AGO, I launched a business in a garage that would grow to become an international firm called Axon Enterprise. You may know that company better by its previous name: TASER International.

Today, over one million TASER weapons have been deployed, saving hundreds of thousands of people from potential death or serious injury. Alongside TASER weapon technology, we led the movement to have police wear body cameras, which have shown great promise to both improve transparency and reduce violent police encounters. We've proven, I believe, that it's possible to address important social problems through a combination of technology and entrepreneurship.

I remain the CEO of Axon, but I am not writing this book in my capacity as a corporate leader. Rather, I am writing it in my personal capacity. My purpose in this book is to challenge conventional thinking about a problem that has plagued human civilization since the beginning: the practice of killing.

I believe we can reduce violence in our world, but we are not doing everything we can to achieve that goal. That's partly because the discussions about public safety, violent crime, and gun ownership are stuck in place. We retreat into online echo chambers where we interact with people who think just like we do, and the battles around pressing challenges become the intellectual equivalent of trench warfare: we dig in, we don't move, we fight for our side, and we yell ever more angrily at those on the other side.

I do not believe that we will solve today's challenges by yelling more emphatically about approaches that have been around for decades or by denigrating people with opposing views. We need new thinking and fresh ideas. We need to explore outside of the echo chambers and interact with people who think differently than we do.

I imagine that many people see me as a pure law-and-order kind of guy. I'm not, and if that's what you're expecting, then I hope this book will disappoint you. On the one hand, I will argue that police need more advanced tools and technology to keep a modern society safe. On the other, each of those technologies carries significant risks of abuse and oppression that must be addressed to realize their promise. We must find the right balance of privacy and security, which is the proper path to both fair and effective law enforcement.

Some of the ideas in this book will seem radical and controversial. I propose, for example, that we must modernize the laws we empower police to enforce, starting with ending the failed war on drugs, which has done little to reduce drug use and has created one of the primary drivers of killing in the modern world. I see opportunities to use artificial intelligence and robotics to change the nature of warfare—not to industrialize killing, but to industrialize *not* killing—enabling military operations without loss of life. I invite activists and academics to engage in helping us envision the problems these new technologies will create, and to design oversight mechanisms to prevent and deter misuse.

My goal in suggesting what sound like far-out ideas is to reach across the intellectual divides to find common ground and provide novel approaches to age-old problems. To accomplish this goal, I have done something that most CEOs of public companies are told never to do: speak my mind freely. This book is a private brainstorm made public. And the last thing you want to do in a good brainstorming session is kill ideas—even crazy-sounding ones—too early. Many times, I've witnessed the ideas we were tempted to write off become the breakthrough solutions we need.

This book includes the thoughts of many people I've interviewed about these subjects, including people who often disagree with me. For you, the reader, I hope this is an opportunity to think critically and creatively. I invite you to imagine how the future might be different. Feel free to challenge me. If you prove that some of my ideas are completely wrong, and do so logically, you will earn my respect and help advance our shared goal of a less violent world.

Suggesting a risky new idea and then having it modified or corrected is an essential part of the creative process. The original idea may not be the precise answer, but it can provide an important stepping stone along the path of innovation. The challenges that are the subject of this book need more free-ranging thinking and rigorous back-and-forth, and a lot less anger and divisiveness. To that end, I invite you to join the conversation on Twitter, posting your ideas, opinions, and refutations with the hashtag #EndOfKilling. Let's all challenge ourselves to keep the conversation civil in an age where that has become ever more difficult, especially on sensitive topics like those covered in this book.

This book includes stories from the development of my company, creating weapons that are designed to incapacitate someone while avoiding death or serious injury. I want to share these stories—some of which I've never shared before—because I think it's important for people to understand what work in this field actually looks and feels like.

Typically, you'll only read about TASER weapons in the press when things go awry. These are incredibly complex and difficult issues, dealing in life and death. I've been through countless lawsuits. I've survived numerous campaigns to discredit me and attempts to end my work. I am pushing hard to change the world, and I have learned through that process that the world pushes back pretty hard. I've been in the "end of killing" business for over two decades. It is my personal mission, and I believe in it, but I've also faced my fair share of critics and naysayers. This book doesn't run away from any of those criticisms.

But this isn't primarily a story about my company or what we've built. Changes to public safety and military technology are happening all around us. How we police our neighborhoods, how we fight our wars, how we pursue justice, how we protect ourselves—each of these is going through a monumental shift. In the same way that the smartphone has reconfigured our relationship with our friends, our music, and our reading, the technologies that are possible today should force us to rethink everything we thought we knew about surveillance, privacy, violence, and killing.

Many of the things I discuss in this book are neither the views of nor the programs of the company of which I am CEO. In that role, I have a talented team of people who challenge my ideas and put them through a rigorous risk assessment before they ever get approved or implemented. But this book needed to be more than just the safe ideas that a public company can entertain in a press release. If I only shared the safe ideas, the book would lose some of its power to drive a change in our thinking. And the opportunity for some crazy idea that ultimately changes the world for the better might have been lost.

Every technology discussed in this book carries a risk of abuse. It also brings a promise of improving safety and enhancing quality of life. I'll touch on the risks, but I will focus more attention on the benefits of these new technologies, largely because I believe the risks receive ample media attention already. We need a more robust discussion of how new technologies can benefit us.

There is a temptation to look at every issue in the world of military and policing only through the lens of George Orwell. In that view, any and all public safety technology—any progress in the tools given to police and soldiers—inevitably becomes a tool of totalitarian regimes. I believe we should also consider the future envisioned by Gene Roddenberry, creator of *Star Trek*. We should consider a vision of a future in which technology elevates humanity and ask how we can guide ourselves toward these more optimistic possibilities.

Join me with an open mind in imagining a different future. These ideas might at first appear scary or bizarre, but beneath the surface we just might find a world with far less human tragedy and suffering, and far more peace and compassion. I am confident that some of these ideas are not yet fully formed—and I invite you to challenge me. You may convince me the risks outweigh the benefits, or you might have the critical idea that tilts the balance toward a better world. And together, we just might find a way to use new technologies to solve some of our oldest problems.

RICK SMITH
Scottsdale, AZ · 2019

PROLOGUE
DRAGONFLIES OVER RAQQA

Raqqa, Syria: April 5, 2045, 2:30 a.m. local time

THE TRANQUIL SOUNDS of chirping crickets and the occasional brush of leaves in the breeze contrast with the scent rising from the courtyard at the center of this small cluster of homes. A putrid smell of decaying human flesh hovers over the city.

A few days earlier, a familiar scene had played out here. Groups of men arrived in pickup trucks, faces concealed behind headscarves and carrying large black banners. Although armed with light firearms and homemade explosives, they used long knives as they went about inflicting violence upon the local inhabitants.

Remnants of the local police force and militia scattered as the newcomers arrived, and those who failed to flee fast enough were quickly rounded up and gruesomely executed in the town square. Their remains—disfigured, dismembered, and mounted for public display—fill the night air with the smell of death.

The town had witnessed this scene countless times over the past decades, as one violent group after another descended upon the inhabitants to assert their dominance. Each attack began with violence and ended with the installation of repressive laws and codes of conduct—violations of which led to more death and dismemberment.

But on this otherwise peaceful night, the villagers hear something different. Through the air, a subtle new sound emerges: a faint hum that seems to come from all directions. Above, patches of the starry night sky go dark for a moment, then the stars return.

Hundreds of metal cylinders with short, stubby wings pass overhead. At first glance, one might presume a very large squadron of fighter jets are passing over the city at low altitudes. Only they seem to be flying much more slowly and quietly. And there are no explosions or gunfire, just the hum of hundreds of low-flying drone aircraft, which look like the cruise missiles that have become all too familiar in conflicts in this region.

As they pass overhead, bay doors open and small, buzzing objects fall out by the thousands. The few villagers who happen to be awake have become so accustomed to violence that they brace themselves for it, waiting to see where and when the explosions will begin. But unlike prior times when aircraft flew overhead and bombs dropped below, this time there are no explosions. There are no bright flashes or buildings crumbling around them. All the villagers can tell is that a swarm of buzzing machines briefly hovered over their homes and then seemed to disappear entirely. Then the quiet night returned, undisturbed.

AS THE sun breaks above the horizon, the city begins to stir. A door opens and a man stumbles out. Covered in blood and dust, with his hands tied behind his back, he is unable to catch himself as he falls with a sickening smack face-first onto the ground. A second man emerges, his face covered with a pashmina and his hands carrying a Kalashnikov assault rifle. He steps forward, grabs the first man by the hair, and violently pulls him upward. He marches the man across the street and toward a telephone pole near the edge of what was once a quiet park. He binds the man's hands to metal rings attached to the pole and draws a large knife, which glints in the sun.

Unbeknownst to either the attacker or the victim, every move they make is being watched. Approximately a hundred feet away, perched atop the corner of a two-story building, a high-resolution 64K sensor is trained upon the activity in the square below. The ultra-high-resolution video stream can distinguish the dust and hair follicles matted in blood, and even the fine movements of the facial muscles of the man being bound to the pole. The video image is converted into a 7G signal and beamed skyward in a one-terabit-per-second data stream that is received by a low-Earth satellite. This is the third generation of the SpaceX Starlink system, one of three major constellations providing high-speed data coverage on every inch of the habitable Earth.

Thousands of miles away, in a large office space just north of Las Vegas, Nevada, an analyst watches the activity closely. He isn't the only one: 1,500 such analysts are watching similar feeds coming from similar sensors scattered thousands of miles across the Earth.

Just a few hours earlier, a fleet of low-speed autonomous aircraft had released 120,000 small drones—known as Dragonflies—over Raqqa. Each video feed was analyzed by onboard AI algorithms seeking signs of human activity. Once triggered, the local feed was prioritized for analysis and beamed to the center in Las Vegas, where more powerful servers crunched through the feeds to identify the most interesting video streams. Those that merited further analysis were then live-streamed to human analysts, with less than five hundred milliseconds latency. A combination of high-power AI algorithms and human analysts were watching and tracking all the human activity in a city thousands of miles away, even as most of the residents of that city were sound asleep.

The villagers watching the scene on the ground have become so used to seeing the inevitable result—a beheading—that they don't flinch or move. They simply wait for the gruesome conclusion. What little movement does take place is parents shielding their children and covering their eyes.

Then, in a split second, everything changes. Thousands of miles away, a human operator gives the "go" command with the click of a mouse. For all of the automation and local machine intelligence, the final decision to use force remains in the hands of human operators capable of making moral choices. The video sensor sitting atop a roofline on the corner of the square pops into the air. The sensor sits on a small quad-copter-style Dragonfly drone. Executing a 180-degree barrel roll, it accelerates toward the target at a speed approaching one hundred miles per hour. Less than two seconds later, just as the blade hovers above the victim, the man holding the knife hears something that causes him to turn his head.

The Dragonfly targeting system selects two locations for maximum effect and calculates three sets of trajectories simultaneously: the speed, direction, and acceleration of the Dragonfly itself, and then the speed and direction of the two darts that it fires in rapid succession over a few milliseconds.

The first dart hits the knife-wielding man at the base of his neck. The second hits his left thigh, at the mid-point of the hamstring muscle. The Dragonfly chose the locations to maximize the effectiveness of its pulsed electric currents. With one dart near the top of the spinal column and another in the middle of the large muscle mass of the legs, the current is sure to run through the entire spinal column—and that means the man's peripheral motor nervous system will shut down. All of the nerves that control arm and leg movements begin in the spinal cord, so voluntary muscle control will be gone. In less than a tenth of a second, the subject is paralyzed and helpless. His blade drops to the ground.

Just as the man catches a glimpse of the quad-copter drone, he goes rigid, experiencing what looks to the villagers like a severe seizure. He falls directly to the ground as the small crowd of onlookers go slack-jawed, stunned at what has just happened. The Dragonfly spins around and lands about three feet from the subject's head, with one video sensor positioned for clear visibility of the target and its 360-degree camera searching in all directions for incoming threats. After roughly five seconds, the man starts to move again; the Dragonfly has cut the electric current off.

Then the device speaks, in a perfect dialect of the local language: "Do not move. You are being detained for human rights abuses. If you move, you will be shocked again immediately." Almost on cue, the subject begins to extend his hands, reaching toward the knife. Then, his arms go rigid again as the Dragonfly applies another few seconds of stimulus. It repeats the warning: "Do not move. You are being detained for human rights abuses. If you move, you will be shocked immediately."

This time, he complies. Confused and disoriented, he understands this much: further movement is a ticket to more discomfort. It has now been about fifteen seconds since Dragonfly One first leapt from the rooftop. A curious young boy in the crowd looks up to try to figure out where the mysterious robotic creature has come from. He notices the sun glinting off similar-looking creatures perched on rooftops in every direction.

By now, another man has emerged from the doorway and is running toward his comrade on the ground. He holds a weapon, and it is clear that he intends to attack whatever strange device has disabled his friend. But he doesn't stand a chance: in a few seconds, he is on the ground as well, stunned by another device, Dragonfly Two, that seems to come out of nowhere.

The second subject is less compliant than the first had been; even after repeated warnings, he is still moving, his mind crying defiance. After the third shock, Dragonfly Two escalates to the secondary protocol, firing a small dart into the muscles of the back, administering a dose of ketamine, a tranquilizer that leaves the subject unconscious within a few minutes.

Directional speakers on the Dragonfly units allow them to direct commands to each subject and to simultaneously send instructions to the growing crowd of onlookers. By now, several additional Dragonfly units have taken sentry positions near the first two, occasionally flying patterns to make it clear that bystanders need to stay back.

Within a few minutes, there is a distinctive sound as a much larger octo-copter drone flies into view. It has been dubbed the Kraken by its creators, in part due to its eight engines, reminiscent of the eight legs of the mythical sea monster, and in part for the long, spindly tentacle it deploys as it does its work. The tentacle is a lengthy mechanical arm, made entirely of metal but with a series of rubberized joints that allows it to move with dexterity and speed. It looks like a giant version of a drain snake—except that it's attached to a drone.

The Kraken comes to hover about twenty feet above the first subject. The tentacle comes down and a small robotic device at its end rapidly secures a harness around the upper torso of the first man, who is lying prone on the ground.

Then, as quickly as it appeared, the Kraken accelerates into the air, lifting the confused man along with it and disappearing into the sky. Dragonfly One ejects the wires connected to the darts and returns to its perch up on the rooftop, settling in right next to the solar panel array it had transported the evening before during its descent. It begins recharging its batteries and resumes observation, as the other Dragonfly units do the same.

Across the city, similar scenes play out. As the day wears on, a number of the fighters determine it isn't safe to go outside and begin to hole up inside buildings. For these situations, small Ground Robots, or Gro-Bots, are ferried in by Kraken drones. The GroBots have been designed to rapidly open doors and climb stairs, and each has a Dragonfly attached that can quickly deploy to augment the GroBots' capabilities.

Finally, a friendlier series of Shepherd robots fan out across the streets, instructing residents to move toward predefined exit points from the city. Designed with large screens that display avatars of local people, the Shepherds provide instructions and guide everyone to safe zones. Anyone who threatens the growing stream of civilian traffic quickly experiences a Dragonfly sting and then a Kraken flight.

About ten miles outside of town, Britain's Royal Marines, a member of the United Nations–led coalition, has set up a detainment and screening center. The first Kraken arrives after a short eight-minute flight with the first subject. He is lowered into what is effectively a shipping container with smooth metal walls and a large bulletproof glass window. The Kraken releases the harness and then flies off for a quick battery swap to prepare for its next mission.

The subject is rapidly screened, disarmed as necessary, removed from the receiving container (with help from Dragonflies and non-lethally armed human guards), handcuffed, and searched. He is then taken to an air-conditioned trailer, where he is given some water and food, medically checked for injuries, and placed in a chair with his arms temporarily bound.

A technician steps forward and straps what looks like a thin motorcycle helmet to the man's head. Similar to those used by video gamers around the world, the helmet contains a high-resolution neural scanning device. Gone are the days of waterboarding, sleep deprivation, or even intensive interrogation. Instead, the technician scans the activity in the subject's brain. An investigator asks a few mild questions to focus his attention on the types of activities the system wants to highlight within the subject's memories. Inside the helmet, infrared light passes through the detainee's brain, and onboard processors analyze the patterns into which the light scatters—a next-generation neuro scanner.

Within a few moments, the system downloads high-resolution memories of several gruesome killings the man has committed. It also evaluates his level of intent and agency and his level of participation. By any estimation, this is a homicidal criminal, and as a result, he will immediately be sent to a high-security detention center. There, he'll await a fast criminal trial and, once convicted, lifelong detention in a military prison.

Over the course of this operation, thousands of people will be processed in the same manner. Some of them will be determined to have done nothing wrong; they will be released and given ten thousand dollars along with a sincere apology. Some will be found to have been coerced, acting as hostages or carrying out acts of violence under threats to their families. Such individuals will be assigned to therapeutic treatments as appropriate.

All the while, the ecosystem of devices will be collecting data, correlating information from hundreds of thousands of live video feeds with the information gleaned from scanned subjects. The system continuously learns to better distinguish between civilian and terrorist. Relevant information is transmitted into highly encrypted systems far from the city's limits and irrelevant personal information is deleted. Rapid neural scans take place privately and as humanely as possible, with all the information sorted in a microcosm of the last century's courtroom practices. The system's designers know full well that the practice of scanning everyone exiting the city risks creating more hostility if people are humiliated by the experience.

However, ensuring everyone's safety requires screening the exodus to ensure violent offenders cannot simply meld into the flow of humanity and escape only to terrorize again. Once scanned, the villagers' facial features and irises are logged for future identification, to avoid unnecessary double screenings and to separate the innocent from the perpetrators more quickly. Similarly, the faces of people in their memories who had performed gruesome acts are logged into wanted lists and uploaded to all robotic systems to assist in rapid threat assessments and capture missions.

Only in the final stages of this military operation will the city's residents see a once-familiar sight: local soldiers and police moving through the city as children return to playing in the parks with a growing sense of safety. These guardians are lightly armed and much more comfortable in their environment than past soldiers; they wear limited body armor, with no need for the clunky military helmets of days past. They have the look of peacekeepers rather than invaders. They perform final checks in close coordination with the various drones and robotic systems that had done prior sweeps. There won't be any shots fired; all the shooters have already been apprehended or neutralized.

Following the rise of ISIS back in the 2010s, the world had learned some crucial lessons. With the right technology, violent insurgencies could be quelled with a minimum of bloodshed; indeed, the best response to an insurgency was one that created no bloodshed. Perpetrators of heinous acts could be separated from the civilian populations they targeted at an early stage. Cycles of violence could be broken before they began. The new United Nation's Peacekeeping and Intervention Policy of 2042 called for fast, highly automated threat mitigation in order to break such cycles of violence before they reach critical mass. And technology played a central role in making it all possible.

INTRODUCTION

KILLING IS A technology problem. And we can use technology to end killing as we know it.

The Raqqa scene that opened the book is an exercise in science fiction, but it illustrates what future armed conflict may actually look like. As futuristic and far out as that Raqqa scene seems, many of the capabilities and technologies described in the story already exist today. It isn't too much of a stretch to imagine that we could fight current and future wars and keep the peace without a drop of spilled blood (or at least with a greatly reduced need for it).

The same goes for the police shooting that leaves a community traumatized and deeply divided: that's something we have the technology to prevent. Awful headlines about school shootings can become a thing of the past. All of that can happen through a thoughtful application of modern technology.

This idea, that new technology can help eliminate humanity's historic violent tendencies, forms the basis of this book. And I suspect I know what you're thinking: this is crazy. Radical. Controversial. Unthinkable. Yes. Yes. Yes—and yes. It is all of those things—but it is not impossible.

As with all truly radical ideas, I expect that this idea will be greeted with disbelief or even ridicule the first time it's heard. And probably the second and third times, too. But to me, this is more than a crazy idea. Building technologies to reduce violence and protect human life is my life's work. I've spent my career trying to create products that make killing obsolete. And I believe that those technologies hold a remarkable, still untapped potential.

I've also come to understand, from experience, that killing is something other than a crime, a sin, or a social ill. It is a problem that can and must be solved. And with the right technology, we can solve it.

YOU HAVE probably heard of the TASER weapon, a stun gun device that uses electricity to incapacitate a human subject. If you're in law enforcement, you might know it because you carry one on your hip. If you're in the general public, you might know it because it has become part of popular culture. Maybe you've seen TASER weapons used in film, and I bet you still remember the "Don't TASE me, bro" meme (selected as the most famous phrase of 2007).

Though you probably know my company's product, you might not know its origin story. When I was in college, two of my friends were shot to death in a parking lot, after a minor traffic accident they were involved in led to a fatal confrontation. It was a senseless tragedy, and also the most pivotal event of my life. Once the problem of gun violence hit close to home, I began to think about it more seriously and creatively.

Gun violence—and violence more generally—has been conceived of as a legal problem, an economic problem, a cultural problem, a moral problem, and more. Through all of those conceptions, it has remained intractable. What if we framed the problem in an entirely new way: as a technology problem? As I asked that question after my friends were killed, inspiration came from a familiar place: science fiction.

I used to devour science fiction as a kid. In *Star Trek*, there's a weapon called the phaser. I'm sure you can picture Captain Kirk (or Captain Picard, if you're a bit younger than me) wielding one, issuing orders like, "Set phasers to stun." I wasn't the first person to look at utopian science fiction and think about how to make it a reality. Others have watched *Star Trek* or similar shows and asked, "Why can't we have warp travel? Or hover boards? Or a peaceful planetary government? Or a 'post-scarcity' economy?" I watched, and asked a different question: *Why can't we have phasers?* Meaning, what if weapons had a setting that would allow you to incapacitate someone rather than end their life?

Sometimes naivete can be the soil of creativity. My naive question—*Could we build a* Star Trek *phaser?*—led me to find a retired NASA scientist who'd had a similar thought thirty years earlier. He and I worked together to make the TASER product, which led to the company I have spent my life building.

The idea at the heart of this book has animated my career: we should not accept killing as an immutable part of human society. We can make the practice of killing obsolete in our lifetimes. Killing is, in fact, a technology problem.

That assertion merits some explanation. First, my primary work concentrates on sanctioned killing—self-defense, police shootings, warfare. In other words, instances in which it is legally permissible to take a life. As technology changes, our standards of what is acceptable and lawful must change as well. Killing that we today consider acceptable and lawful is precisely the kind of killing we can be rid of, and sooner than you might think, once we have the right technology to make it obsolete.

Second, I will explore strategies to reduce murder and violent crimes, the illegal killings and predation. Some of these approaches are policy oriented—such as modifying or eliminating laws that underpin efforts like the "war on drugs," an effort that that, I believe, perpetuates violence. If the primary role of government is to ensure a safe and peaceful society, then when we identify laws that serve as engines driving violent behaviors, we should take swift action to address or eject them.

Simultaneously, there is an opportunity to leverage the growing network of cameras and sensors to ensure violent crimes are effectively prosecuted or even prevented. In every incident of violent crime, the perpetrator is placing a bet that they will not get caught. Modern technology can decrease the probability of getting away with it, resulting in fewer criminals taking the bet and those who do being rapidly removed from society so that they cannot offend again.

More broadly, I believe modern weapons carried by soldiers and police are antiquated, just bad upgrades on the musket. If we invest our resources, scientific know-how, and political willpower in upgrading the technology we use, we can make killing every bit as outdated. We could build weapons that don't destroy human life, but rather incapacitate and subdue threats. In cases in which sanctioned violence needs to occur—a police officer immobilizing a suspect or a soldier capturing an enemy—we can develop effective alternatives to killing.

THESE ARE ideas, but these ideas aren't abstract and they have life and death consequences. Almost 40,000 people a year are killed with firearms in the United States, and 250,000 are killed annually worldwide. A new era of weapons technology could bring that number down. In the process, it could change how we think, talk, and legislate the business of killing.

There will still, unfortunately and tragically, be gun deaths and gun crimes. The world is a violent place, and this book is not about the end of violence or the end of death, but the end of killing. I don't want to pretend that technology can fix all of what human beings do to one another. That kind of techno-utopianism has, at times, proven either misguided or dangerous. At the same time, I don't want to sell the thesis short. I think lives can be saved if we embrace the idea that a lot of modern killing happens because we haven't thought hard enough about the tools being used to police our streets, fight our wars, and protect our homes.

Killing is not some hard truth of human nature that we are powerless to stop. For thousands of years, killing was the primary means for determining the outcome of many interpersonal or intercommunity conflicts. Thankfully, the amount of killing in the world is dropping dramatically. Killing is no longer accepted, except in limited edge cases where we have no other suitable technological choice. This book is about creating those choices and enabling technologies and policies that further the historic decline in violence, until killing is just a bad memory we read about in history classes.

WE NEED robust public discussions about the effects of technology, about how we use new devices, and about how we legislate their impact. The smartphone, for example, has been a great boon, but it's also forced parents to give new thought to the relationship between their kids and technology. Cloud technology has been an incredible way to store and manage terabytes of data, but it has also challenged all of us to think hard about privacy, data protection, and the amount of our information that entities outside of our direct control possess.

Right now, weapons technology is undergoing a similarly epochal change—but we haven't had an open and meaningful discussion about it. Companies are currently developing the next generation of lethal and non-lethal tools, but the laws that govern our weaponry were made for an early era of weapons. More information about us is available to the government and private sector, but few people understand how that information is being used and even fewer know the rules that govern that use. How are our laws, norms, and expectations changing to keep pace with these transformations? Extend this thought exercise to include ubiquitous cameras and the surveillance they allow, and the always-uneasy balance between old laws and new tech should lead us to think hard (and fast!) about the law and norms for a future that is quickly arriving.

In some cases, that future is already here. In the professions that my work most closely touches, these questions come up every day. For example, if a body camera is attached to a police officer's chest, how does that change their behavior? How much or how little of that information should be allowed in a courtroom? If a soldier carries a weapon that's designed to incapacitate rather than kill a target, does it make them more trigger-happy or less? Should people be allowed to keep *Star Trek* phasers at home, in the same way that they can keep guns at home? If drone technology could be sent into a foreign city to gather intelligence house-to-house and capture suspected terrorists, what laws and international covenants should govern the use of that technology?

It's tempting to think that there are faraway experts that should determine the answers to these questions, a special council, priesthood, or tribunal. But there aren't. We—the voting and non-voting public—are the ones who need to figure out the answers. The public not only needs to be informed about the choices that are being made, but we also need to play a role in shaping and making those choices. Police officers are responsible for public safety, meaning your safety and mine. The decisions they make and the work they do affects all of us. So their jobs and the way they go about them are public questions.

This is a very nuanced topic, and unfortunately, nuance is often the first thing to go in the modern media. The media has one simple objective: to get your attention. Therefore, negative stories that trigger people's fears dominate the pages of your local newspaper and favorite news sites. The newspaper adage "If it bleeds, it leads" is truer in the digital age than it's ever been before.

I speak from experience: my company and I have been heavily criticized in the press. I've been critiqued by news outlets, faced off against the National Rifle Association (NRA), and been targeted by various activist groups. Over time, I have learned not to take it personally. Fear sells. Our prehistoric brains are highly attuned to dangers and threats. The humans who survived and passed on their genes were those whose ancestors were attuned to danger, threats, anxiety, and fear. In a world where we usually no longer have to fear that a lion will eat us for lunch, those fear-attuned genes are still with us. So the media companies give us exactly what we respond to: fear and animosity, lots of it.

The media has led many to believe the world is getting more and more dangerous, when exactly the opposite is true. By almost any measure, the human condition has improved dramatically over the past few centuries. Life expectancies have more than doubled, literacy rates have exploded, and extreme poverty has plummeted. The risk of death from war or violence has fallen by over 90 percent (see www.EndOfKilling.com/progress).

There are certainly risks in the future, and some of those risks carry huge costs. But those risks shouldn't cause us to give up; in fact, they should do the opposite: they should give us hope and the courage to continue to improve the human condition.

OFTEN, AS WITH the opening story set in Raqqa, I'll illustrate my points about the future of killing with scenarios that represent what will be possible in the near future. In most cases, I draw from weapons technology that already exists or is in early stages of development.

Why borrow from the tools of science fiction for a non-fiction book? Because I want to imagine what might not be possible this instant but could be possible in the not-so-distant future. This technology and its uses can be complicated to explain, and I want to paint clear pictures of what wars, conflict situations, and police actions of the future could be like. I want you to understand these ideas in a way that's visceral and visual, not just intellectual. It's important for those who don't deal in this business every day to understand what it means, and you won't do that just by digesting data or watching the news. Stories have the power to shake us out of our received notions and rigid habits of thought, to expand our sense of the possible.

The book addresses the different forms and places in which the weapons technology of the future will change how we fight, shoot, kill, arrest, and police. I've spoken to leaders in the field, interviewed a wide range of technologists and military leaders, and come up with a sense of what the next era of this work might look like.

Public and personal safety, homeland security, and national defense will all be irrevocably changed by technology, but it's important for citizens to understand how it's changing, to ask hard questions, and to take an active, engaged role in the process of social change that always accompanies technological change. Big issues about privacy, security, and liberty are at the heart of this transformation. Many of these questions do not have easy answers; many of the answers will challenge our received wisdom, our sense of what's reasonable, and even of what's regarded as good. But we can shape those answers with an eye to making the world safer in the process.

We should all challenge the assumption that killing will forever be part of the story of humanity. Rather than be frozen in the face of that complexity, let's dive in and test, let's try to implement. Not acting without precautions, but not being overcautious either.

To end justifiable gun violence and sanctioned killing in the next few decades, we will need clear and ambitious goals. And in that spirit, this book is animated by four audacious goals, which I believe we can achieve within a reasonable timeframe:

1 Policing without killing—by 2030
2 Military operations without bloodshed
3 Big tech companies helping reduce violence
4 Activists who advance progress in police and military institutions

POLICING WITHOUT KILLING—BY 2030

The last decade has been, arguably, the hardest period in the history of police work. Deaths in the line of duty remain a constant risk for every officer. Never has there been more scrutiny on what police officers do, how they do it, and why they do it. Activists have gone after police in the public, the press, and the streets.

We lament police shootings, and they generate significant energy and emotion within communities. Anger and frustration tends to focus on officers' racism or the malicious intent of those involved in the shootings—and in some cases, those criticisms are right. However, there is too little focus on how we might be able to avoid repeating these problems by rethinking the underlying tool sets and how we can systematically improve outcomes.

It is easy, but intellectually lazy, to label all police officers as racist. That alienates an entire segment of society comprising about one million Americans (and another ten million globally) who have dedicated themselves to a career in public safety. The law of large numbers applies here, as it does in most areas of life: in any sizable group of people, there are no doubt a few people in policing's ranks who hold prejudiced views. But just because *some* people who hold those views are police officers, that isn't reason enough to paint all police officers with a broad brush of prejudice.

Besides, it doesn't help matters to apply sweeping labels: progress in any field doesn't come from name-calling, stoking anger, and brewing resentment. It comes from thoughtful discourse, invention, and experimentation. And from using hard data rather than raw emotion to measure outcomes and drive better results.

To be a cop in the United States in 2019 can feel like you're perpetually under siege, that you're the bad guy, not the guy protecting others *from* the bad guy. Police officers speak to me every day about what they're up against. They're desperate to restore police work to its admired place in society. They want to be appreciated and respected; they'd rather be loved than feared.

No police officer enters the force looking to kill. And by 2030, I don't think any police officer should have to kill. In order for them to protect communities, and themselves, they will need access to weapons that immediately suppress a threat more effectively than the lethal weapons they rely upon today. By 2030, I believe that the most effective tools for mitigating threats will no longer need to take lives. Alternatives will be accurate, immediate, and more effective than aiming a handgun at a suspect and firing off a round, hoping that you hit your target.

This mission is not a critique of the men and women who do the job of public safety with the lethal force tools they must rely upon today. When we imagine the technologies we need to develop to send humans to Mars (and bring them back again), we aren't criticizing the astronauts of today who haven't gone there yet. Similarly, when we imagine the tools that will make policing without killing a future reality, we aren't criticizing the officers who go to work today with current tools. Instead, I want to motivate the scientists, politicians, activists, technologists, and leaders who can help create the tools to make that future better than our present.

MILITARY OPERATIONS WITHOUT BLOODSHED

Imagine a World War II commander hearing this proposition: two of the five largest land armies on Earth are amassed for a fight. One side has had years to dig in. Both are armed with tens of thousands of soldiers, thousands of tanks, aircraft, artillery, and all the accoutrements of modern warfare. One side attacks . . . and wins complete and utter military victory with close to zero combat casualties.

The World War II commander would be astonished by a war of that kind. And yet, this is exactly the technological and military accomplishment demonstrated in the first Gulf War, enabled entirely by the superior technology of one side. And that's one of the cornerstones of this book: the idea of war without combat deaths isn't as far-fetched as you might think it is at first blush.

In fact, a sure sign of human progress is that deaths from armed conflicts have gone down dramatically over the past centuries. There are a variety of reasons for that, including the end of superpower clashes; treaties that restrict the use of certain kinds of weapons; and the success, so far, of nuclear deterrence. But there's one big blind spot in our national security apparatus that could drive those numbers down even further: the incredible potential of non-lethal weapons to carry out missions more effectively.

Notice I said "carry out missions more effectively." Not "stop protecting lives and property and ideals." Not "stop intervening in cases of genocide or ethnic cleansing." I believe police and militaries can carry out their missions *and* do so without taking life. And I stress that the technology I'm talking about doesn't mean compromising security for even one second. What I'd like policymakers and the public to understand and explore is the idea that non-lethal options can actually help the military *better* achieve its mission objectives.

Today, the lack of effective non-lethal military weapons is a strategic Achilles' heel for armed forces around the world. Our inability to stop a child approaching a military vehicle without them being killed is a weakness that adversaries exploit with brutal creativity. Baiting soldiers into killing innocents has become a powerful strategic weapon.

Through interviews with a range of military thinkers and leaders, I've come to learn how badly soldiers need options that don't require the taking of life to stop a potential threat. I've also come to understand how much intelligence can be lost with a bullet; in many cases, we've killed enemy combatants who would have been treasure troves of information. We gave that information up because there were no effective non-lethal options at our disposal. Furthermore, we continually put soldiers in unwinnable situations because of a lack of available options.

One of the people I work with most closely experienced this firsthand. He was a Marine officer in Iraq. While commanding a checkpoint, he and his troops faced the almost impossible decision of what to do as an ambulance approached them at high speed. They tried signaling for the driver to stop; he didn't. They finally tried their best to fire precision shots at the tires and engine block. The vehicle stopped and the Marines moved closer to inspect what had happened.

Inside they found that they had gravely injured a pregnant woman. They tried life-saving procedures, but it was too late. Both the woman and her unborn child died, with the Marines helpless as the entire ghastly scene played out in front of them. No Marine signs up for military service to kill pregnant women. That is not why they decided to serve their country. We must ask: Surely in an era in which humanity has created the technology for cars to drive themselves, we could find a safer, less lethal way to deal with a car driving up to a checkpoint?

First, we must imagine that the end of killing is possible and that achieving it is a goal worthy of pursuit. Then, we must change our thinking to invest in a new direction. Consider this: the plan to modernize the United States' nuclear weapons arsenal, first proposed under the Obama administration, calls for spending $1.2 trillion on next-generation nuclear weapons. By comparison, the investment in non-lethal capabilities rounds to zero. If we continue to say that lethal weapons are the only choice we have, that's at least in part because we've chosen to make them the only choice.

This is, in many ways, a problem of perspective and principles. As Marine Colonel Scott Buran, a leading researcher and teacher on non-lethal weapons, told me: "The present military mindset is locked into lethality and kinetic operations . . . It's been an up-hill battle to get the military to change that mindset." For him, it's not so much about the tactics—the actual technology that makes non-lethals work—as it is about the buy-in of the military and their civilian leaders into the idea of alternatives to lethal firearms and weaponry. In other words, this is a strategic and philosophical issue, not an operational one. "We have to get beyond the tactical discussion of non-lethals," he said. "We've got to move into the ethical, the philosophical, and even the theological to move people's hearts and minds. It's hard enough to change their minds. But you will have to change their hearts."

What we need, and what I believe we can bring about, is exactly this shift in both hearts and minds. It will be difficult to change an institution as big and tradition-bound as the military. But rapidly evolving threats require fresh solutions, and the seeming impossibility of solving a problem is no excuse for not trying.

BIG TECH COMPANIES HELPING REDUCE VIOLENCE

Big tech companies are at the forefront of the information age and the digital revolution. These companies draw some of the brightest minds of our generation. And yet they've walled themselves off from solving some of the world's most pressing problems.

Google, among others, has decreed that it will not allow the military to use its technology. Period. Facebook has proclaimed that its response to gun violence is to block ads for any and all weapons manufacturers.

That's it? That's what the biggest and most-talked-about companies are doing to address the pressing problems of our time? Banning contracts and blocking ads?

Once you become a publicly traded company—once you possess the size, reach, market cap, and talent that these companies have—you also carry an obligation to put that talent to use against problems that cry out for answers, not just problems that are fun, happy, comfortable, or profitable.

Imagine, if you will, Google deciding that a crack team of its engineers would work on the problem of school shootings. Or Facebook deciding that it was going to use its data and artificial intelligence capabilities to bring down crime rates in a neighborhood. Not in the spirit of reluctant cooperation with the government, but as a leader of the pack, a company that believes that its reserves of talent and wealth could be put to the most difficult challenges our society faces.

At the moment, what these and other big tech players have decided is that rather than do the hard work of rolling up their sleeves and fixing the problems of violence and crime, they will let someone else do it. This is a serious dereliction of duty by some of the most talented men and women of our time. We need our brightest minds on our toughest problems. Can we make school shootings a thing of the past? Can we eliminate death from warfare? We don't have concrete solutions to those questions, but I know that it's going to take all our talent to answer them. And that talent often resides in Silicon Valley companies that have chosen to look the other way, often (somewhat ironically) claiming a moral high ground in the press for doing so.

ACTIVISTS WHO ADVANCE PROGRESS IN POLICE AND MILITARY INSTITUTIONS

For my entire career, I've worked (and often struggled) with activist groups. The American Civil Liberties Union (ACLU), Amnesty International, Human Rights Watch—I've tangled with them all. I've also taken a drubbing from the powerful NRA lobbyists, who successfully blocked the airlines from using non-lethal weapons so that guns would be the only approved option in the 2002 legislation that armed airline pilots in the wake of the September 11 attacks.

These discussions are often about finding the right balance between very different worldviews, which is why we need serious discourse between different stakeholders to map out the future. Even as they've sometimes been resistant to new approaches, I've learned a lot from these groups. We don't always agree, but they are as committed to their causes and points of view as I am to mine.

That said, I've also seen activist groups evolve in ways that have become counterproductive and actually work against the causes they stand for. It's a common pitfall that people will become defined by— even consumed by—what they are against rather than what they are for.

What they stand against broadens until it becomes unrecognizable and even counterproductive. Being against violent police encounters, for instance, can degenerate into being against police work altogether. Being anti-war can lead you to become anti-soldier. People will rarely admit when this happens, but it's human nature to allow one lens through which we see part of a problem to become the only lens through which we see all problems.

Thus groups that seek to reduce excessive force in policing can end up resisting any changes in policing, even changes that would help reduce injuries and stymie the use of force by police. I am speaking here from personal experience. Independent studies have repeatedly shown that TASER weapons reduced the number of people injured by police. But because TASER weapons were perceived as a new type of force (and indeed, because it was any type of force), they were immediately decried by some activists. Those groups came out strongly against this new type of force without sufficient regard to what it could actually do on the ground to reduce adverse outcomes.

In the most productive cases, activists engage to drive toward improved outcomes. In chapter 12, I will describe the collaboration between the ACLU, the Cincinnati Police Department, and my company that resulted in a dramatic reduction in police shootings.

Unfortunately, sometimes activist groups thwart the very progress they'd like to bring about in the world. International human rights organizations, for example, have been actively protesting the use of any non-lethal technology in warfare. They argue that having access to non-lethal weapons would make warfare more likely. In other words, because weapons would be available that would make killing less frequent, nation-states might be more likely to use weapons, period.

The idea that non-lethal technology would increase the propensity to conflict simply doesn't make sense to me, and no study confirms or supports this view. But the objections of international human rights organizations still carry weight in the public and press, and many a non-lethal project has been scuttled because an activist group opposed it.

That's a problem. It obstructs progress, and it prevents us from saving lives. I think activist groups can be far more effective if they work constructively with governments and private sector entities to implement new solutions and strategies.

IT IS VERY EASY when discussing these topics to get caught in the "good" versus "bad" debate. I grew up playing cops and robbers, a game rooted in the idea that there were good guys and bad guys. But as I have grown up, I have learned that life is much more complex than a child's game. Sure, there are some real-life bad guys. And there are some brutal cops. But it's far more common that there are tragic situations. For example, people struggling with mental health crises or drug intoxication that expresses itself in a violent and deadly outburst. Police officers are affected by the same things all human beings are, and they can crack under the pressure of seeing a friend or colleague killed. They can find themselves in a murky situation where they make a bad decision—or even lose control and make a vindictive one.

Those decisions and difficult choices carry enormous consequences, well after the conflict ends. How many soldiers, sailors, airmen and airwomen, and Marines have come back from the bloodshed in Iraq and Afghanistan with moral injuries because of what they've endured? And how many of those things have been done because there weren't better, less lethal, more precise munitions available to achieve their objectives?

These aren't hypothetical questions. They are the everyday, real struggles of people who are charged with carrying out violence in our name. These people deserve better than they've been given. They also deserve a public that is better informed about the choices they are forced to make. I believe that once the public understands these issues, they will become exasperated by the fact that we don't already have better options available.

History, I believe, is moving in the direction of peace—toward a world in which killing is abnormal, rare, and unnecessary. In 1945, the American military demonstrated the awesome and destructive power of nuclear weapons; never in history had a weapon been designed with more lethality. And yet, less than four decades later, in 1983, President Ronald Reagan called on scientists to solve the problem their research had created: "To turn their great talents now to the cause of mankind and world peace, to give us the means of rendering those nuclear weapons impotent and obsolete."

His proposal—the ill-fated "Star Wars" plan to intercept missiles in mid-air with an elaborate ballistic shield—didn't go far. But the idea at the heart of his plea, that technology and the people who build it could reduce killing and not increase it, has been carried forward by a new generation of innovators, whose goal has been to make weapons of all kinds "impotent and obsolete." Success today is no longer a simple and indiscriminate increase in force; there's no longer a prize for a bigger "boom." In fact, the goal of modern weapons is the precise opposite of their predecessors: to minimize the number of deaths and collateral damage. To eliminate the threat without eliminating the life.

If this book paints an optimistic portrait of the future—good. We have reason to be optimistic. Even if the headlines blare with reports of crime and war, there is a quiet revolution taking place in weaponry, one that could make state-sanctioned killing a thing of the past. But we need to encourage those changes, discuss the ramifications of new technologies, and work to build the future we'd like our children to live in.

We're at a powerful moment in the history of weaponry. I think we can, within our lifetimes, turn guns into museum pieces and relics of the past. And in so doing, we can bring sanctioned killing to a decisive and necessary end. The story of violence is as old as the story of humanity. We're about to write a new chapter in it.

WEAPONS,
PAST AND PRESENT

ON THE SAVANNA of Senegal in 2007, scientists observed chimpanzees wielding what is perhaps the oldest primate weapon, one that had been previously thought unique to humans: the spear. The chimpanzees' target was the galago, or bush baby, a small, nocturnal primate that spends its days curled up in the hollows of trees. Galagos are easily startled, and chimpanzees who reach directly into their nests will usually find their prey scampering quickly beyond their reach. So, over thousands of years of trial and error, the chimpanzees have learned to use a tool in pursuit of their meal. Gripping a sharp stick, they quietly approach the tree and then repeatedly stab their spear into the hollow. If they're lucky, they extract a freshly killed galago, pre-skewered and ready to eat.

Killing with a spear is—as it has been for all of primate history—brutal, bloody work. The separation between the killer and the killed is just the length of the weapon. Killing with a spear means directly confronting your victim: watching blood pour from the wound, feeling the pressure of the body on the spear's shaft, watching the light go out of their eyes. Perhaps none of that is troubling for chimpanzees killing other, smaller primates; we have little way of knowing. But we do know that killing, for the vast bulk of human history, meant dealing with just those sorts of sensations. Killing was something we did face to face, with our hands.

The history of our human weapons is, in one sense, a story of how control of nature's raw materials granted some humans power over animals and over other humans. A good wooden spear, whittled down to a sharp point and hardened in the fire, can pierce a cave bear's hide. A flint-tipped spear is stronger than a wooden spear. Bronze is stronger than flint, iron is stronger than bronze, and so on. Humans with access to superior weapons technology have depopulated entire continents of their megafauna, defended their tribes from raiding bands, and built kingdoms and empires. Human history is an arms race; the same process that began hundreds of thousands of years ago, when we first learned how to use tools and first discovered their power for taking life, still goes on in countries around the world.

That's one way of telling the story of weapons. Let me propose another way of telling that story—as a story about visibility. The earliest weapons were much like that spear wielded by the Senegal chimpanzees, a means of face-to-face killing, in which the victim, and the act of killing itself, is immediately visible to the killer (with all of the potential psychological costs that implies). Taken in the broad sense, advances in weaponry have separated killer and victim more and more, so that killing no longer needs to be a face-to-face act. The pilot of an F-35 need never see or know anything about the individual humans on whom their plane's bombs fall.

There's another, related way in which killing has become less visible. There was a time when powerful regimes flaunted their ability to inflict violence and death. The visibility of killing was itself a source of their legitimacy. This, too, has changed over time. States still kill, of course, but whereas earlier governments deliberately made killing as visible as possible, today's governments strive whenever possible to hide it. They use drones and targeted assassination, both because they are more effective and because they keep killing out of public view. Hand-in-hand with this development is the more recent evolution of laws of warfare and conventions for regulating state violence—yet more checks on what states are willing to be seen doing.

In this chapter, we'll explore both of these developments: the ways in which technology has diminished the prevalence of face-to-face, hands-on killing, and the ways in which states have used technology to shield organized killing from view. It's possible to be cynical about these developments, to claim that we've just gotten better at hiding our killing, and that chimpanzees, whatever else we might say about them, are at least forced to face up to the consequences of their violence when they stab galagos to death. But that's not the view I take. The decreasing visibility of killing certainly hasn't been driven by noble motives. And yet it has the effect of making humans, in general, more squeamish about killing, in a beneficial way. When members of a society are less and less willing to kill with their own hands, and when the government of that society is less and less willing to advertise its organized killing to the world, that society is on the road to treating killing itself as unacceptable.

I STARTED this brief history with spears for a reason: spears are older than modern humans. As long as Homo sapiens have existed on Earth, we've wielded spears. Given the prehistory of weapons is an inexact science, and though the dates may be revised as more archeological evidence becomes available, it seems that the earliest evidence of spear use dates to 400,000 years ago. At that point in the distant past, modern humans had yet to branch off from Homo erectus and other early hominids. In other words, modern humans inherited our earliest weapons from our hominid forebears. As an article in the journal *Nature* put it, announcing the discovery of the remains of those early spears alongside stone tools and the remains of butchered horses, the evidence suggests "that systematic hunting, involving foresight, planning and the use of appropriate technology, was part of the behavioral repertoire of pre-modern hominids." Along with the earliest weapons, then, we inherited from those ancestors a capacity for organized, efficient killing. Other animals are capable of coordinated hunting, as well; think of a wolf pack, for instance. But early humans' ability to coordinate at a larger scale, and with the aid of increasingly sophisticated lethal tools, gave them an edge over their competitors on the ancient savanna.

Perhaps modern humans' first unique contribution to the art of organized killing was the atlatl, the first tool that enabled killing at a distance. It has been called the Stone Age Kalashnikov, a nickname that testifies to its lethality and ubiquity. The atlatl, whose invention dates somewhere between thirty thousand to forty thousand years ago, is a spear-thrower: it consists of a grip at one end and a notch at the other, in which a spear or other projectile is placed. With skill and practice, the atlatl user can gain leverage and power over the spear's trajectory, propelling it over distances of up to three hundred feet at a top speed of nearly a hundred miles per hour. Of course, the identity of the inventor or inventors of this remarkable weapon were lost to time—but we do know that, with its creation, modern humans were able for the first time to attack their prey from a considerable distance, gaining added safety and the advantage of surprise.

Perhaps most remarkably, the atlatl became a truly globalized technology: remains of spear-throwers have been found from Europe to Alaska to Australia. And just as importantly for our story, it changed the experience of killing. A skilled atlatl user attacked from a safe distance and then came upon a victim who was already dead or dying. Already, humans had found ways to remove themselves from the visceral act of killing.

The bow and arrow worked on much the same principle, allowing humans to impart far more force and velocity to a projectile than mere muscle power would allow. The bow and arrow ultimately displaced the spear-thrower. They enabled greater distance and accuracy, and they could be produced in greater numbers: the physics of the atlatl dictates that stone spearheads generally must be customized to their spear-thrower, whereas stone arrowheads were compatible with any bow. These arrowheads, the earliest of which date to eight thousand to ten thousand years ago, are our earliest evidence of bow and arrow use. But another early piece of evidence is even more striking: the oldest known depiction of combat, found among Mesolithic cave paintings in Valencia, Spain, shows a battle of archers, in which four attackers surround a huddled group of three defenders. While the earliest human weapons may have been developed with hunting in mind, humans had clearly learned to turn them on one another.

The next great revolution in weaponry was connected to innovations in metallurgy. Some four thousand years ago, humans discovered that copper combined with tin formed a new alloy that was tougher and longer lasting than any previously known metal: bronze. This discovery was so instrumental to the formation of the earliest states that archeologists have long designated the period from the discovery of bronze to around 1000 BCE as the Bronze Age. Bronze Age technology included more durable and more lethal spears, arrowheads, and swords, along with chariots that functioned as mobile weapons platforms.

One the one hand, bronze swords made killing just as visceral and close-quarters an act as it was for the early, spear-carrying hominids. But on the other hand, the organizational capacity of the Bronze Age states turned killing from the work of small, face-to-face war bands into the work of large-scale armies, obeying orders, organized with a greater degree of bureaucratic rigor and bearing mass-produced weapons. Homer's *Iliad* offers a vivid description of two Bronze Age armies on the cusp of meeting in battle:

> Now when they were marshalled, the several companies with their captains, the Trojans came on with clamour and with a cry like birds, even as the clamour of cranes ariseth before the face of heaven, when they flee from wintry storms and measureless rain, and with clamour fly toward the streams of Ocean, bearing slaughter and death to Pigmy men, and in the early dawn they offer evil battle. But the Achaeans came on in silence, breathing fury, eager at heart to bear aid each man to his fellow.

The Iliad brings to mind the exploits of warriors like the fierce Achilles and the doomed Hector. But remember that these warriors, and their historical equivalents, were backed up by masses of men like those in the passage just above. Such men may have obeyed their captains or gone into battle eager to aid their fellow soldiers, but the men on the other side of the battlefield were faceless to them. This was itself a new development in the history of killing. In the same way, Iron Age innovations made possible even more durable weapons technology, but the most successful Iron Age armies, from the Spartan hoplites to the

Roman legions, were first and foremost triumphs of organization and discipline—that is, of depersonalized killing.

A world away, alchemists in China were developing the technology that would once again revolutionize killing: gunpowder. Invented as early as the ninth century CE, gunpowder was famously put to a host of other uses—from medicine to fireworks—before its lethal power was realized. But by 1132, soldiers of the Song dynasty had used the first proto-firearms in battle. These "fire lances" were modified spears, whose tips were set alight before battle—more a psychological advantage than a powerful new technology. About a century later, Chinese soldiers carried "hand cannons" into battle, the first proper handheld firearms.

By the fourteenth century, gunpowder technology had diffused as far as Europe and India. Early handheld firearms were still dangerous to their users, unwieldy to aim, and difficult to load, so it was as siege weapons that firearms made their most immediate impact. That impact was felt, for instance, at the gates of Constantinople, where a Turkish army under Mehmed II used its massive siege cannons to sack the city, bringing to an end the last remnant of the Roman empire. Meanwhile, in Europe, engineers were perfecting the harquebus, the first triggered firearm (which resembled a small musket). With the aid of similar weapons, European armies finally brought an end to the military usefulness of the mounted and armored knight, who had carried Iron Age technology into the early modern era.

As firearms grew increasingly sophisticated—with the widespread adoption of rifling for greater accuracy in the nineteenth century, and the first machine guns in the 1850s and '60s—killing on a truly industrial scale became possible. These developments reached their peak with the invention of the first fully automated firearm in 1884: the Maxim gun, which was used to devastating effect in Africa by soldiers of the British Empire. In a way, the Maxim gun looks back to the earliest Homo sapiens weapon, the atlatl: it is an extremely advanced way of exerting violence at a distance, and separating the killer from the act, and the consequences, of killing. As human weapons technology has advanced, this separation has grown increasingly pronounced. Before we reflect on the consequences of this separation, though, I want to observe the other way in which organized killing has become less visible.

ONE OF THE most famous public executions was inflicted on Robert-François Damiens in 1757. He had tried and failed to assassinate King Louis XV of France, and as punishment he was sentenced to death by torture. His execution—the last one by drawing-and-quartering in the history of France—fascinated a number of philosophers and social critics, among them Michel Foucault. Quoting contemporary sources, Foucault showed how the public execution was choreographed in excruciating detail:

> Damiens ... was to be "taken and conveyed in a cart, wearing nothing but a shirt, holding a torch of burning wax weighing two pounds ... to the Place de Grève, where, on a scaffold that will be erected there, the flesh will be torn from his breasts, arms, thighs and calves with red-hot pincers, his right hand, holding the knife with which he committed the said parricide, burnt with sulphur, and, on those places where the flesh will be torn away, poured molten lead, boiling oil, burning resin, wax and sulphur melted together and then his body drawn and quartered by four horses and his limbs and body consumed by fire, reduced to ashes and his ashes thrown to the winds."

Damiens's execution, argued Foucault, was remarkable not just for its brutality, but also for its publicity. For centuries, this was how states and kings demonstrated their power—by publicly and proudly inflicting violence in front of a willing audience. By attempting to take the life of the king, Damiens had struck at the heart of the social order, and so the social order had to demonstrate that it was still intact by paying that violence back in front of the biggest crowd possible. Execution was a spectacle.

While the methods employed in eighteenth-century France may have been especially gruesome, the principle was an ancient and venerable one: rulers demonstrated their power and legitimacy by killing, publicly and visibly. The government of Louis XV showed it was still in control by killing a would-be assassin for all to see; other kings personally advertised their valor in battle. One of the most famous depictions of an Egyptian pharaoh, for instance, shows Ramses II mowing down his foes with bow and arrow from a chariot. Thousands of years later, warrior kings like Gustavus Adolphus of Sweden still had themselves painted leading men into battle, in full military regalia. Dealing death, publicly and visibly, was what sovereigns did.

And yet, the other remarkable fact about Damiens's execution was that it was the last of its kind; it marked, Foucault wrote, "the disappearance of torture as a public spectacle." Consider another execution, which took place in 2018. On the night of February 22, the State of Alabama attempted for more than two hours to execute Doyle Lee Hamm. Hamm had been convicted of murder and sentenced to death in 1987, but by the time his appeals finally ran out, cancer had seriously damaged his veins and made execution by lethal injection—the state's prescribed method—virtually impossible. Prison officials punctured his skin eleven times in search of a usable vein and were unable to find one; they finally declared the execution postponed. In March 2018, Alabama agreed to forego any further attempts to execute Hamm, and he remains in prison.

I bring up this case as a contrast to Damiens's because Hamm was saved by the modern state's squeamishness about visible killing. For the most part, death-penalty states insist on death by lethal injection, which takes place in private, behind high prison walls, and is designed to resemble a medical procedure as much as possible. Instead of trying to put Hamm to sleep on a hospital gurney, the State of Alabama might have easily killed him in any number of ways, including one as simple as a bullet to the head. But those methods have the disadvantage of looking like killing.

Centuries ago, sovereigns had no concerns about acts that looked like killing—they wanted to be seen killing as often, and as theatrically, as possible. Today, though, states are so wary of visible killing that even their executions are designed to look like doctor's appointments.

Or think of it this way: Who wants organized killing to be more visible today? Invariably, it's people who think that there is too much organized killing happening. Take police body cameras, for instance. Imagine them teleported back into the Middle Ages, and you'd likely see kings gleefully strapping them on sheriffs and constables and broadcasting the results, even (or especially) when those sheriffs and constables killed offenders. In those days, visible killing bolstered a king's power. Today, though, the people who push hardest for police body cameras are those who believe police shoot suspects far too often. They want to make killing visible so that there will be less of it.

Or consider warfare: far from having themselves depicted slaughtering their enemies with their own hands, in the manner of Ramses II, today's states describe warfare using terms like "precision-guided munitions" and "surgical strikes." The iconic image of the first Gulf War is a night-vision shot of a "smart bomb" being deposited down an Iraqi chimney, the government-endorsed message being that modern technology takes all of the messy unpredictability out of war. Conquering armies are no longer awash in glory through devastating the enemy; now they broadcast their own casualties in order to generate sympathy in the court of public opinion.

I think we can see the evolution of laws of war as part of this broad development. Some basic laws of war—such as respect for heralds and envoys, and truces to bury the dead—date back to antiquity. But the modern effort to codify laws of war took shape with the Hague and Geneva conventions, which attempted to codify definitions of war crimes and regulate treatment of non-combatants. The Hague Conventions of 1899 and 1907 dealt with conduct in war: they attempted to ban such tactics as aerial bombardment from balloons, the use of automatic submarine mines, and the deployment of poison gas on the battlefield. In 1925, an addition to the conventions banned all chemical and biological weapons. The Geneva Conventions—a series of international treaties originating in 1864 and updated most recently in 1949—were intended to govern actions toward civilians and others not engaged in combat. In addition to setting standards for humane treatment of prisoners of war, these conventions outlaw torture and attempt to protect civilians from armed conflict.

As you read that series of dates, you may notice that the histories of the Hague and Geneva conventions are marked by flagrant violations, from the mustard gas of World War I to the concentration camps of World War II. The sad fact is that regulating warfare has proved elusive; nations are only subject to the laws they choose to obey, and time and again they have chosen to break international law when it suited them. But if there's a hopeful way to understand that history, it's this: even when they are not willing to abide by laws of war, hundreds of nations have still gone on record endorsing and supporting them. Even if their conduct has been hypocritical, hypocrisy is, after all, the tribute that vice pays to virtue. It is no longer acceptable to be seen breaking the rules of war or exercising wanton violence. That is significant.

To sum up: developments in technology, and in the perception of violence, have conspired to gradually remove humans from the act of killing in a range of cases. Pushing a button to drop a bomb is different from impaling your enemy on a spear. Tearing a criminal's body apart in public is different from injecting a criminal with heart-stopping drugs on an operating table. Killing is now less immediate to those who do it, and arguably less visceral to the public that watches it.

And maybe you're asking: So what? A bomb kills every bit as much as a spear does—probably more effectively. An execution is an execution. Perhaps we've gotten better at hiding killing, but that doesn't mean we're approaching the end of killing.

Maybe not. I certainly don't think many of the developments covered in this chapter happened out of the goodness of people's hearts. Early Homo sapiens didn't invent the atlatl because they thought it was a more humane way of downing an antelope; they invented it because it worked better. And yet our inventions can have, over the centuries, unintended consequences. The less we see killing, the less we're likely to accept it as a normal part of life. The less a state broadcasts its use of violence, the less likely we are to see violence as the way a state builds its legitimacy. Today, I think that humanity as a whole is far more squeamish about killing than at any point in our history. While it's true we consume violent content in films, television, and video games, the idea of watching a public execution in our local community is beyond the pale. We see killing as abnormal, disturbing, outrageous—the exception rather than the rule.

None of that means that an end to killing is inevitable. But it does mean that, unwittingly, we've been building up a psychological and political foundation for an end to killing. Up to this point, the technological revolutions in weaponry have distanced us from killing. I believe that the next technological revolution will make the end of killing not just conceivable, but also achievable.

THE NEW VS. THE NOW

OW DO WE develop non-lethal weaponry that's safe, effective, and 100 percent reliable? How do we use artificial intelligence and big data to prevent violence proactively, rather than reacting to violence after it occurs? How can technology help us immobilize and disarm threats, rather than killing now and asking questions later?

The technical challenges are real, and overcoming them will take considerable applications of ingenuity, creativity, and persistence. (See www.EndOfKilling.com for more detail on those technical challenges.) But in my view, none of those challenges is the main impediment to progress. The main impediment is a psychological challenge. It lies in our deep-rooted aversion to "the new."

When I talk about our aversion to the new, you might already be registering an objection. Isn't our whole economy built around convincing people to spend money on newness? Don't businesses, from tech companies to auto manufacturers to fashion designers, roll out new iterations of their products every year, if not multiple times a year? Don't we all respond to the lure of "new and improved"?

Sure. But let's recognize the difference between superficially new and transformatively new. The next iPhone may have a few more bells and whistles than the current iPhone—but fundamentally, it's still an iPhone. The gap between two models of the same smartphone is

39

nothing compared to the gap between a world in which no smartphones exist and a world in which, one day, they do. The latter gap illustrates the kind of transformative newness that I'm talking about. And it pales next to even more transformative change—for instance, when our cars suddenly start driving themselves.

There are plenty of good reasons why humans avoid the new. One is evolutionary. It's well known that our risk-averse ancestors were the ones who lived and passed on their genes, not the risk-acceptant ones. In human prehistory, the costs of taking a risk and getting it wrong—death and the potential end of your genetic line—far outweighed the marginal rewards of taking a risk and getting it right. If many of us are cautious by nature and quick to pounce on the faults and drawbacks of a course of action, that's because millions of years of natural selection have strongly favored those traits.

Today, psychologists and behavioral economists study those traits as they manifest themselves in phenomena like "loss aversion." Loss aversion, which was famously studied by Amos Tversky and Daniel Kahneman, refers to our tendency to weigh potential losses much more heavily than potential gains. Study after study of people making economic decisions bears this out: we prefer avoiding a loss to acquiring a gain, even when the probabilities and amounts are equivalent on both sides. In strict economic terms, protecting a hundred dollars that you already have in the bank is the same as gaining an additional hundred dollars. But in behavioral terms, most of us will value keeping the money we have in the bank much, much more highly (according to some studies, almost twice as highly). This is a sort of inborn conservatism that affects most of our decisions: we'd rather avoid losing what we have than trying to gain more.

Combined with these individual factors biasing us against the new are some powerful social factors. Every incumbent way of doing things—from legacy technologies to old ways of organizing a business—has a powerful army of defenders just because it is incumbent. Powerful and successful people have an interest in preserving the systems that made them powerful and successful: buggy makers had an interest in the failure of cars, and oil companies have an interest in the failure of electric vehicles.

Of course, "creative destruction" does happen. But it happens less than it might, because the incumbents in danger of destruction generally use all of their power and resources to fight back. This resistance often gets magnified by a media industry that is fighting for attention and needs scary headlines to get it, by interest groups that also need to grab that attention to stay relevant, and indeed, by our own brains, which are hyper-attentive to new risks.

Our evolutionary legacy of caution, loss aversion, and the power of incumbent systems means that every transformative technology must overcome an enormous set of hurdles. These hurdles often appear in the mental tests we devise for new technologies, tests that are subtly slanted toward failure. I'm sure you've heard the saying, "The perfect is the enemy of the good." It's not just an aphorism; it's actually how we think about transformative change. We don't compare a proposed change to the current world and ask if it will be, on balance, an improvement. We're much more likely to compare that proposed change to perfection, and if it comes up short, to turn against it.

Think, for instance, of the enormous media coverage generated in 2018 when a self-driving Uber SUV struck and killed a pedestrian in Tempe, Arizona. It goes without saying that the death was a tragedy. But keep in mind that around the world, an average of 3,287 deaths happen on the road each day—3,287 tragedies that generate only a fraction of the attention directed to a single death in Arizona.

The question implied by that disproportionate attention is clear. It's not "Would a world of self-driving cars be safer than the current world of 3,287 car crash fatalities per day?" It's "Can self-driving cars achieve a perfect record of avoiding fatalities?" Until they do, we'll have a powerful bias against them, not because they are unsafe, but because they are new. New technology is forced, in other words, to meet a higher standard than the technology of today. What's new must be "perfect," not "better."

When it comes to new technology, we see this same pattern again and again. In the same way, cases of battery fires in Tesla's electric cars generated a great deal of negative attention, sparking fears that these vehicles are fundamentally unsafe. It didn't seem to matter that gas-powered cars can catch on fire after accidents, too. In 2013, Elon

Musk wrote a convincing piece about the comparative risks of fires in electric and gas-powered cars; while Musk obviously has a stake in the issue, he's not making up the statistics. The numbers speak for themselves. As he put it:

> There are now substantially more than the 19,000 [Tesla] Model S vehicles on the road that were reported in our Q3 shareholder letter for an average of one fire per at least 6,333 cars, compared to the rate for gasoline vehicles of one fire per 1,350 cars. By this metric, you are more than four and a half times more likely to experience a fire in a gasoline car than a Model S! Considering the odds in the absolute, you are more likely to be struck by lightning in your lifetime than experience even a non-injurious fire in a Tesla.
>
> Those metrics tell only part of the story. The far more deadly nature of a gasoline car fire deserves to be re-emphasized. Since the Model S went into production mid last year, there have been over 400 deaths and 1,200 serious injuries in the United States alone due to gasoline car fires, compared to zero deaths and zero injuries due to Tesla fires anywhere in the world.
>
> There is a real, physical reason for this: a gasoline tank has 10 times more combustion energy than our battery pack.

And yet, the question posed here—"Are Teslas safer than gasoline-powered vehicles?"—isn't the one that dominates coverage of these issues. The question that usually gets asked isn't, "Are Teslas safer than the status quo?" The implied standard is, "Are Teslas perfectly safe?" As I've suggested, that question itself reveals cognitive bias, because we never subject the status quo to the same rigorous standard.

Of course, the media has powerful incentives to play into these tendencies. The media also has powerful incentives to play up surprising news. An entirely new kind of car crash is more attention-generating than the kind of car crash that happens every day. More generally, news, media, and entertainment all thrive on clash and conflict. A sinister new technology with deadly side-effects is the kind of conflict that propels a story forward, the kind of conflict I'm sure we've all seen many times on the big screen. A dispassionate weighing of risks—balancing the risks of the status quo against the risks of change—might be valuable, but it is hardly ever interesting in the same way.

To help understand the risks of new policing technologies, both real and perceived, my company, along with several privacy and police oversight experts, recently created an AI and Policing Technology Ethics Board. Before our first full meeting, I assigned some homework: I asked everyone to watch the movie *Minority Report* so we could discuss it over dinner the night before our meeting.

In many of the more sensational press stories about new policing technologies, there are references to how these new technologies might bring about the "dystopian world" of *Minority Report*. So I thought it would be an entertaining and productive exercise to rewatch the movie and to use it as a basis for discussion.

In brief, the technology in the movie took a country that was being overwhelmed with murders and drove the rate of murders down to *zero*. This was accomplished through a new technology called "precognition." Putting aside the sci-fi handwaving that makes precognition a reality in the movie's world, let's focus on the results.

The system of precognition had a perfect record of predicting and preventing murders. Then the main plot twist unfolds—the man who had created the pre-crimes program murdered someone and, using his unique position and knowledge, he was able to circumvent the system and get away with it.

So by the end of the movie, we see the effectiveness and fairness record drop from 100 percent to something like 99.99 percent due to one error (again, from the one person in a unique position to influence the system). As a result, the movie ends with the termination of the entire pre-crime program, and all previous convicts are released to roam the streets. And the audience cheers, with a sense of relief, that justice was finally served.

Movies can make us do amazing things. They can get us to root for a serial killer and cannibal like Hannibal Lecter, as he kills and mutilates prison guards to escape from jail disguised as a corpse. They can get us rooting for Walter White, the meth-dealing high school teacher from *Breaking Bad*. The power of story and identification with protagonists can overwhelm our rationality. And in the case of *Minority Report*, the fear of the new is ultimately justified with a big payoff moment.

But let's compare the world of *Minority Report* to the world we live in (and the world that would have existed before the technology in the film and presumably would return after the program is shut down). The real world is beset with violence and murders. The pre-crime program drove murders to zero. Let's talk about justice. The pre-crime program had an accuracy rate of 99.99 percent. Does anyone believe that the justice system in the real world is anywhere close to that accurate?

In fact, the people who are quickest to decry anything with a whiff of *Minority Report* in policing are the same people who are most critical of the justice system as it now stands. The world in which we live delivers justice based on a system of ritualistic performances in front of an audience of twelve jurors. Twelve ordinary people get it wrong sometimes; of course, the racial bias in our society creeps into our justice system as well.

My point isn't that the *Minority Report* pre-crime system is ready and waiting to be rolled out tomorrow. It's science fiction. My point is that fiction shapes the way we think about the real world, and in this respect, *Minority Report* is highly relevant and highly representative: it imagines a system that is better across the board than what we have now, finds an imperfection that leads to an injustice, and then tells us to be afraid of that system because it's not perfect. And we applaud for that fear.

With every new policing technology that is proposed—from facial recognition to predictive policing to drones—we see endless references to *Minority Report* and George Orwell. And by contrast, we see far less discussion of how these technologies might be used responsibly to improve both community safety and justice over the often haphazard systems of today.

As we discussed the film over dinner, one of the guests ironically repeated a quote that is often attributed to Joseph Stalin: "One death is a tragedy; one million deaths is a statistic." It's a powerful statement, and painfully true. One tragic death in a self-driving car sets back progress against a backdrop of the status quo where thousands of deaths occur every day from humans driving.

When it comes to evaluating new technologies for their costs and benefits, we might expect the media and pop culture to get it wrong, and if we're smart, we might turn to more reliable guides. But the power of loss aversion goes even deeper than that, because it shapes the thinking of institutions we might expect to weigh costs and benefits more rationally.

One of those is the military. Later on (see chapter 7), I'll tell in more detail the story of the Active Denial System (ADS), a non-lethal weapon that can use a focused "heat ray" to disperse hostile crowds or deter enemy forces without indiscriminate killing—an especially valuable feature when so many hostile insurgents are difficult to distinguish from the civilians they hide among. It was fear of the new that killed the Active Denial System and hampered the US Military's development of non-lethal weaponry. The status quo bias in favor of the "tried and true" solutions of bullets and bombs has, ultimately, made the military's task in conflict zones even more difficult, because insurgencies thrive when our forces wrongly target civilians with lethal weaponry. I'd argue that the military has turned away from the promise of non-lethal weapons, because, not unlike the media, it's asking the wrong questions: not "Is a mistake committed with a non-lethal weapon less serious than a mistake committed with a gun?" but rather "Are the lives saved from using non-lethal weapons worth the bad press we will get from deploying something new and scary?"

We can even see this kind of thinking in the cars we drive today. Did you know that automotive airbags were invented as early as 1951? And yet it took about four decades for them to become widely adopted—in part because the technology took time to develop, but also because accidental or harmful airbag deployments received a disproportionate amount of attention and because for years major automakers successfully lobbied against airbag requirements. Those automakers would turn to members of Congress and emphasize those accidental and harmful deployments, and we missed the opportunity to measure the risk of requiring airbags compared to the risk of going on without them.

We're wired to treat the new as scary. It's a fundamental premise of this book that this is an irrational way to think about the world. Some new things are scary. Some new things aren't. What makes them legitimately scary isn't their newness, but rather their degree of risk compared with the status quo. Every time we consider the benefits and drawbacks of a new way of approaching a problem, we should not ask ourselves if the new way is free of drawbacks. It never is. We need to ask ourselves if it's better than what we have right now. Asking that simple question requires overcoming deep-seated cognitive biases and generations of baked-in loss aversion. But it's possible, and with practice, it gets easier.

As you read this book, I'm going to ask you to confront those biases in yourself, and to try hard to overcome them. Many of the concepts in this book will feel strange, even creepy. But we need to step back and compare these future possibilities to the grotesqueries of the world we live in. Cops shooting unarmed kids. People being sentenced unjustly based on imperfect information and human prejudice. Countless civilian casualties across the globe, in communities where militaries are trying to help root out bad actors by dropping five-hundred-pound bombs from the sky.

We spend far too much time focusing on what could go wrong, and not nearly enough time on what might go right. I'm not asking you to be credulous, to stop asking hard questions about new ways of doing things. But I'm asking you to ask the same hard questions of the world you live in right now. Don't weigh the new against the perfect. Weigh it against the now. Evaluate the new against the world of today, assessing where some smart and informed risk-taking can make the world, on balance, a less violent place.

WHY WE NEED
TO STOP KILLING

UNDER WHAT circumstances are police trained to shoot to kill? The answer may surprise you: Never. Police are not trained to shoot to kill. Ever. The legal and moral doctrine that allows police officers in the United States to discharge a weapon is to "shoot to stop the threat." Police are not expected to shoot with the objective of killing a person; they are expected to neutralize threats to themselves or to third parties.

What about the case of a police sniper who is explicitly authorized to kill a hostage taker with a well-placed shot to the head? The reason that action is authorized is not for the purpose of killing the target. The action is authorized because a bullet shot to the brain is the most reliable method to immediately stop the threat.

These two objectives—which are very different in theory—become conflated in practice, to the point that many people assume police are trained to shoot to kill. There's a clear reason why: currently, killing by way of firearm is the most reliable means of stopping a threat. The former police commissioner of Philadelphia, Chuck Ramsey, who, in 2014, headed a White House task force on the future of policing, gave his succinct definition of the justified use of force: "not using anything beyond what is absolutely necessary to stop the threat." When an officer uses force, what he's ultimately looking for, as Ramsey told me, is "time to process. Time to get the situation under control and resolve it, short of someone dying . . . what you're looking for is a window of time—five seconds, ten seconds, maybe twenty—whatever it takes to approach and safely take someone into custody."

Here's an analogy that may help. Until the invention of reliable modern anesthesia in the nineteenth century, surgeons often encouraged patients to drink copious amounts of alcohol to numb their pain. Those surgeons' objective wasn't to get patients drunk; it was to minimize the pain of an incision or an amputation. But when medical technology advanced, surgeons could do away with the means to an end represented by a bottle of whiskey or rum. The end didn't change; the means grew more refined.

The next time you read about a police shooting, look for these words from the officer: "I had no choice." You will always see them, in some form or variant. And in most circumstances and in most courts of law, it's an acceptable rationale that assumes killing was the only choice available to stop the threat. But it also highlights that killing is an unintended consequence, the unfortunate side-effect. My goal is to examine what happens when that link is broken and what combinations of technology can break that link.

In other words, what happens when the most reliable and effective technology to stop a threat is a non-lethal force? If a non-lethal option can more reliably and quickly incapacitate a threat, then under what circumstances, if any, would a deadly option be justified? I argue that it essentially would never be justified. Chief Ramsey agrees. When I asked him when he would authorize the use of deadly force if non-lethal and lethal options had the same effectiveness, his response was blunt: "I wouldn't. There would be no need." He added that there is a reason why police officers are trained to shoot for the center of mass— the middle of the body—because it's the place they are most likely to be able to stop a threat.

A bullet can never be taken back. Given the existence of reliable non-lethal weaponry, it is hard to imagine a scenario in which the irreversible option of lethal force would be preferred. When non-lethal force performs better than lethal force, killing becomes unjustifiable. And when that happens, a huge burden will be lifted from those who today have to make life-and-death decisions.

THE PRIMARY resistance to this point is a disbelief that the technology is possible. But disbelief is often the precursor to a breakthrough.

Consider, for example, the Wright brothers. No one believed it was possible to create a device that could fly. And when they successfully invented manned flight, the Wright brothers were all but ignored: the local papers that knew about the story didn't cover it; even the Associated Press turned the story down. When they offered to explain to the US government what they had done, they received a polite form letter declining their offer. But in time, their creation would change humanity. As Albert Einstein put it, "For an idea that first does not seem insane, there is no hope."

An "insane" idea, like manned flight, or space travel, or humanity's ability to end killing in the future, should begin from first principles, with fundamental questions like: Why do we kill at all? At one level, we kill because it's how we survived. It's how our species managed to beat the evolutionary odds and remain a species. It's how we dealt with prehistoric threats, and it's how tribes who felt threatened by each other dealt with their differences. If you look back across history, that all makes a fair amount of sense and it explains much about human civilization.

It's why, for example, our weapons have become more lethal with the passage of time. Technology for killing grew out of the necessity to kill and the better killers a tribe had within it, the more successful it was. Thus the arms race began. Our fists led to clubs; clubs gave way to the bow and arrow, which begat the cannon and rifle; those projectiles laid the groundwork for the bomb and the missile. The ability to maim became, over centuries, an ability to massacre, and at each turning point in that story, the goal was identical: to increase the lethality of the tool and raise the death toll. The person who could exact the biggest toll had the most potential for success.

Up to a certain point, that made sense. At one time—the era in which we were fighting prehistoric creatures—the ability and power to kill were vital. They were essential for survival. "Kill or be killed," the old adage goes, and when facing a hostile and threatening environment, with animals and creatures that saw us as prey, the focus on the ability to kill as a human strength made a great deal of sense. If you're staring at a saber-toothed tiger readying to charge at you and your tribe, you know you have but one option: kill the beast or lose your life.

But in modern times, that logic ceased to hold. In fact, it became absurd. We're no longer facing prehistoric dangers, and the idea that killing is the only answer in law enforcement and military situations is increasingly being questioned. Should we kill? Do we need to? What price do we pay when we take a life, and what are the alternatives?

These can seem like airy, philosophical questions—but they have real-world, brass-tacks implications. Around the country and the world, police departments and militaries wrestle with these questions. We see them play out on the news. A police officer shoots a suspect, a tragic circus ensues. And each time, there's a collective feeling of "Shouldn't we be able to do better than this?"

Killing has a cost, and the price we pay may be bigger than the benefit we get from taking life. To some, that idea will seem obvious; to others, it'll seem like heresy. There are many who believe that simply taking the life of a criminal is preferable to any other outcome. That's an extreme position, but we live in an era in which fringe positions can flirt with the mainstream in unexpected ways. That's why it's important to engage even the most out-there views—they can metastasize more quickly than we think.

So it's worth discussing why we should try to eliminate even premeditated, state-sanctioned killing. I'd submit that there are two big categories of objections to the use of premeditated force: 1) human beings pay an enormous price when they take a life, and that price includes the psychological affliction to the person doing the killing, in addition to the life or lives taken; 2) killing is, ultimately, counterproductive. Both are worth exploring in detail.

THE HUMAN COST OF KILLING

There is obviously a profound human cost of killing. But usually, when we focus on the cost of killing, we focus on the life taken, not on the person taking the life.

But the person killed is only half the story. The other human cost in killing is one that's often neglected: the damage done to the killer in the process of taking a life. "There is no glory in killing anyone," Hans Marrero, a former Marine, said to me. "Every life you take you live with. A certain smell that reminds you of that moment brings back those memories, brings back dreams after that. A noise. An explosion. The smell of blood. You're looking at flies . . . flies always remind me of dead bodies. All that stuff comes back."

Marrero is, by any estimation, one of the most fearsome men on the planet. At one point, he was the Marine Corps hand-to-hand combat instructor and a dedicated warrior. To this day, he remains of the world's experts in how to snap a neck to take someone's life. ("If you train well, it should only take you four or five seconds. Actually, four or five seconds is too long. You can do it sooner.")

Marrero has practiced his trade around the world, in highly classified operations in dangerous places. And this hardened soldier, comfortable with the tools of killing, is one of the leading proponents of a world in which warriors don't have to kill, for the reasons he described above: taking a life doesn't mean that life goes away. It lives on in the mind of the killer.

"My teacher said, 'Be careful when you take a life, because if you take a life, it may change the outcome of the world,'" Marrero said. "If I can avoid killing and give someone a second chance . . ." When Marrero trailed off at this point, it was clear how much thinking he had done about the lives he took in his line of dangerous work. "First time I killed . . . it was for a few seconds, but I sat and I thought about it. For that moment, that guy that I took out, it just caught me off guard. I was the youngest one on the team, so it caught me off guard. And I remember feeling empty inside. It's a horrible feeling. I remember thinking, 'God, does that guy have a mother like my mother? Does he have someone that loves him like my family loves me?' That only lasted like ten seconds . . . but it hit me and dawned on me," Marrero said.

We assume that because our police officers and soldiers are trained in using weapons, they have some kind of impenetrability against the consequences of taking life. And on rare occasions, those views are supported by the stories we hear about a soldier boasting about their kill count or a cop bragging about a shooting they were in. But for Marrero, those kinds of boasts make him suspicious. "If a guy brags too much about killing," Marrero said, "I'm betting it didn't happen." Because, as Marrero has come to understand, the psychological effects of taking that life cast a long shadow. Killing creates a deep wound in the psyche that most people would prefer to leave buried.

Most of the public doesn't understand this notion. In the public's view, if you're trained to do dangerous work, you're somehow magically protected against all of these psychological outcomes. As it turns out, that's not true. You can become proficient in weaponry without being desensitized to killing. And we're beginning to understand that there is great harm done to the person who takes a life, even when their job requires it of them. In fact, for Marrero, it's the soldiers and police officers and others who believe killing is easy and clinical that he worries about. "Killing is not an easy thing to do," he said. "And anybody that says that it's easy, it's because they're a freaking psychopath or sociopath ... He's not right in the head. If you're a good warrior, you don't think that."

We've paid dearly to learn these lessons. In particular, troops returning from Iraq and Afghanistan and getting diagnosed with PTSD (post-traumatic stress disorder) or PTSI (post-traumatic stress injury) have opened up the discussion about what killing does to human beings. The psychiatric casualties of the wars in Iraq and Afghanistan have been as terrible as the physical casualties. The deepest scars from those wars are often the scars we can't see.

Military psychologists are coming to understand that it's not only the typically "traumatic" experiences of wartime that can produce debilitating symptoms in combatants, but also the act of killing itself. That's even the case for drone pilots, who kill from a safe distance in front of computers in air-conditioned rooms. Some of the more striking findings in this area come from Shira Maguen, a psychologist with the San Francisco VA Health Care System (SFVAHCS). She was cited in a 2018 *New York Times Magazine* piece about the subject:

> Dr. Maguen wanted to see if there might be a relationship between taking another life and debilitating consequences like alcohol abuse, relationship problems, outbursts of violence, PTSD. The results were striking: Even when controlling for different experiences in combat,

she found, killing was a "significant, independent predictor of multiple mental health symptoms" and of social dysfunction...

The veterans in Maguen's groups didn't speak much about fear and hyperarousal, emotions linked to PTSD. Mostly, they expressed guilt and self-condemnation. "You feel ashamed of what you did," one said. Others described feeling unworthy of forgiveness and love. The passage of time did little to diminish these moral wounds.

Past generations didn't have the language to talk about these kinds of "moral wounds." But today, we're dealing more squarely with the harsh realities of wartime and our media and films have also changed to reflect the impacts of war.

One of the most popular films about the conflict in Iraq was 2014's *American Sniper*. It was about the Navy SEAL sniper Chris Kyle, someone who was exceptionally well versed in the practice of killing. He had over 160 confirmed kills to his name, and at one point, the enemy placed an eighty-thousand-dollar bounty on his head.

American Sniper is a war drama, and it contains all the action you'd expect from a movie about a Navy SEAL directed by Clint Eastwood. But to his credit, Eastwood also went deeper: at one point, we find a dazed and confused Kyle, returned from a deployment, sitting in a bar mere miles away from his home but unwilling to return to his family for reasons he can't understand. It is one of the more moving and accurate depictions of the "moral wounds" of warfare in modern media.

In my line of work, I've seen police officers speak about these sorts of wounds, too. Officers have told me stories of being left with no option but to take a life, and then revealed how haunted they were by the outcome.

We know now that these aren't just one-off anecdotes or urban legends. Solid research reveals the depressive and post-traumatic effects of taking a life, or even seriously injuring someone in the line of duty. As one such study put it, "Prior research has found that many officers involved in shootings suffer from 'post-shooting trauma'—a form of post-traumatic stress disorder that may include guilt, depression, and even suicidal thoughts." And as documented in 2011 in the *Journal of Psychiatric Research*, "After controlling for demographics and exposure to life threat, killing or seriously injuring someone in the line of duty was significantly associated with PTSD symptoms (p < .01) and marginally associated with depression symptoms (p < .06)."

Psychiatrists confirm what we regularly hear in our business from front-line police officers: taking a life, or grievously injuring someone in the line of duty, carries with it an enormous personal and professional weight. It stays with the police officer for the rest of their days. The images play out over and over again in their mind, and the questions—Was there an alternative? Could something have been done to prevent that incident?—never stop.

SOME FINE work on these issues has been done by Lieutenant Colonel David Grossman. As his military title would suggest, Grossman is no pacifist. He's a former Army officer who has trained men to kill. His family is also in this line of work: his son earned three Bronze Stars for his Air Force service.

His books are popular within law enforcement and military circles, and he is outspoken on these matters and willing to share blunt and hard truths. His 1995 book *On Killing* explores the psychological cost of taking a life. It's required reading by the US Marine Corps commandant, and it's also a text used by the Peace Studies program in Berkeley. That's because it offers a balanced and nuanced look at the act of killing and what it means for those who are entrusted with keeping the public safe.

Grossman reminds us that while our technology for killing has improved with time, human beings have a deep-seated, built-in aversion to taking life. As he put it to me, "Inside most healthy members of our species is a pretty powerful resistance against killing our own kind . . . Healthy people have to be trained to kill." He opens his book with the following, bracing intention: "My prime motivation has been to help pierce the taboo of killing that prevented these men, and many millions like them, from sharing their pain. And then to use the knowledge gained in order to understand first the mechanisms that enable war and then the cause of the current wave of violent crime that is destroying our nation."

Grossman notes that even people who are trained to kill have to fight the hardwired instinct not to. Grossman writes about soldiers in World War II: "Colonel (retired) Albert J. Brown, in Reading, Pennsylvania, exemplifies the kind of response I have consistently received while speaking to veterans' groups. As an infantry platoon leader and company commander in World War II, he observed that 'Squad leaders and platoon sergeants had to move up and down the firing line kicking men to get them to fire. We felt like we were doing good to get two or three men out of a squad to fire.'" The better part of his book is devoted to "the simple and demonstrable fact that there is within most men an intense resistance to killing their fellow man. A resistance so strong that, in many circumstances, soldiers on the battlefield will die before they can overcome it."

Grossman tells us that, again and again, the people best equipped to kill have an overriding psychological reaction that makes killing extraordinarily painful. And when we force police and military personnel to override those circuits, we do lasting and, in some cases, irreparable damage. The decision to kill can ruin someone. That is a cost we must consider in any discussion of the use of premeditated force.

Even though Grossman's work focuses on the psychological aftermath of killing, he's also quick to point out that officers tell him that dealing with the outcry after a killing is just as deep a source of concern for them. "Every cop lives in fear of having to kill someone and being eaten alive by the media," Grossman told me. "I've had cops tell me: 'The killing wasn't the cause of the PTSD. It was dealing with the media that caused the PTSD.' And I believe them 100 percent." Cops speak openly to Grossman about the kind of public shaming they've seen other officers experience. "The aftermath dynamics can make it hard," he said.

Grossman is also witness to some of the more complicated aftermath of what happens when a suspect is crippled, not killed. When a police officer sees a victim testify in court about how they'll be bound to a wheelchair for life because of what the officer did—even if it was justified—that affects them psychologically. In other cases, officers are vilified for not pulling the trigger. All of it creates an environment in which cops are damned if they do, damned if they don't. "Cops can be sued for not shooting ... So here cops have this dilemma about what to do during a deadly force incident," Grossman said. "If they kill 'em, they can be eaten alive by the media. If they cripple 'em, it can be even worse from a psychological perspective."

Whether from the difficulty of a death, the stress of media pressure after a shooting, or the challenge of seeing someone crippled after an officer-involved shooting, these outcomes speak to the fact that the human costs of killing go far beyond the taking of one life. And we need to consider those costs more seriously and more often than we do today.

KILLING IS COUNTERPRODUCTIVE

I think it's important to distinguish between killing as an end and killing as a means. The latter is what matters to law enforcement, the military, and anyone who is in the act of protecting themselves and others. Our soldiers and police officers are trained to shoot with precision, not because killing is the purpose of their work, but because an accurate shot from a firearm is considered to be the most effective and efficient way of neutralizing a threat.

That's an important distinction: if we're going to end sanctioned killing, then it's important to hammer home the point that killing isn't a goal; it's just the most effective way for those our society entrusts with lethal force to handle the threats they encounter today. That also matters because there is real value in controlling and capturing a threat rather than simply pulling the trigger and ending a life.

Again, you might roll your eyes and accept that statement as a given, but there are those who believe that the best enemy is a dead one. They discount the value of capturing and holding a suspect. Military intelligence officials and criminologists around the world may not share the same views on much, but I wager that all of them would argue that someone suspected of wrongdoing can almost always be a powerful source of information. There is value in being able to capture someone, extract intelligence, and then use that intelligence to save lives.

In some cases, better intelligence will make the key difference between winning and losing a conflict. Dr. Gregory Elder, a Defense Intelligence Agency and Central Intelligence Agency researcher, illustrates how, throughout history, comparatively better intelligence can make up for comparatively weaker weaponry. In five consequential battles—the First Battle of Bull Run (1861), Tannenberg (1914), Midway

(1942), Inchon (1950), and the Israeli air strike initiating the Six-Day War (1967)—Elder demonstrates "that it was neither technology nor material superiority that won the day, but accurate, timely, actionable intelligence, combined with leaders willing to treat intelligence as a primary factor in deciding outcomes." In other words, the militaries who won weren't superior killers; they were smarter about gathering and collecting information.

That becomes more important than ever in modern conflicts. As one person I interviewed put it, "The Geneva Conventions were written at a time when adversarial forces were not embedding themselves inside civilian populations and weren't willing to use infants as well as women and children as human shields, which is the very scary case that we see today." In the United States' mission in Afghanistan, it would have been foolhardy—and probably impossible—to find and kill all members of the Taliban. The allegiances were too porous, the enemy too diffuse. Our modern enemy doesn't wear uniforms or abide by ordinary rules of war or Geneva Conventions. They often have an indirect or passing connection to the conflict, and they aren't taking their orders from a general or a political authority.

That means, counterintuitively, that killing them can be deeply counterproductive. Because while they have only weak ties to the conflict itself, they have very strong ties to their communities and tribes. Those connections are often why we read reports of radicalization after the dropping of an errant drone missile or the accidental killing of civilians. This is the perverse irony of many of our modern conflicts: we can't kill our way out of them, and in fact, killing can make the conflict worse, not better. It can make the outcome we'd like less achievable, not more achievable.

As one Taliban official commented to the *New York Times* in early 2019 regarding US Military strikes in Afghanistan, "The more they kill, the more our blood will turn hot."

Better, it would seem, to immobilize the enemy, detain them, and learn what we can from them. Grossman agrees, and he went so far as to criticize the killing that happens by way of drone missile: "If we were able to snatch these guys instead, it would be psychologically vastly more effective than just killing them. Killing has potential in deterrence, but it has potential for empowering them. We honestly don't know at this point whether the Hellfire missile whacking bad guys around the world is going to generate this incredible groundswell of opposition, or whether it is just taking key players out of the equation, or whether it is getting a hold that will lead them to give up on the battle. We just don't know."

It's not just in warfare that killing could be said to be counterproductive. In police work, the same logic holds. Gang networks, drug cartels, and other entities of that kind are brought down by deploying sophisticated intelligence-gathering methods and patiently working a case for a long period of time. Those cases live and die by intelligence, and intelligence is acquired when living suspects can be pressured into giving up sources, names, dates, and other vital information.

That may seem like an obvious point, but it's worth reflecting on it again. A lifeless suspect cannot give you any information about a case or crime. They can't lead you to their compatriots, tell you the location of a hideout or warehouse, or strike a plea bargain in exchange for valuable information. When immobilizing a threat is equated with killing, those options are immediately taken off the table.

KILLING IS NECESSARY—UNTIL IT ISN'T

There are entire libraries dedicated to the study of those two points: that killing carries a high personal cost for the killer, and that the loss of a suspect in a crime can mean a loss of vital intelligence. But on both points, the prime rebuttal is a simple one, which you'll hear from a lot of people who wear uniforms and protect people: "Sometimes there are just no other options."

Time and again, in our work with law enforcement, we hear of a situation—a shootout, a high-speed chase, a hostage scenario—where the only available answer is the taking of a life. "I had to do it," an officer will say. "It was either him or me," another one will tell us. "What other choice did I have?" someone else will add. "I couldn't throw my baton at him."

Those are perfectly reasonable arguments in the heat of the moment, and, I should note, they are almost always delivered with disappointment, exasperation, or regret. I've never—not once—had a police officer who took a life talk to me about how great they felt after the fact. That just doesn't happen. Even in instances where police officers are tried and found guilty of firing without cause, they're devastated by the outcomes and remorseful about their actions. And they almost always believe they had no other choice in the thick of a tense situation.

But what they're describing isn't an affirmative case in *favor* of killing; they're describing a problem that needs to be solved—a problem with the technology we give those officers and the situations in which it's applied. Remember the distinction between killing as a means and killing as an end: the end is to stop another person from shooting, not to stop their heart beating. As long as a criminal is immobilized, whether they're alive or dead has no bearing on whether hostages go free, whether a standoff is resolved, or whether the responding officer gets to go home to their family. Killing, given our present state of technology, may or may not be justified from situation to situation. But killing, in and of itself, is never the goal. The goal is restoring peace, keeping the police and the public safe from danger, or otherwise defusing tough situations.

Given what we know about the cost of killing and taking a life, and what we know about how valuable sources can be in building cases and taking down networks of criminals, what if we had an alternative to killing someone?

That question has dominated my professional life.

4

THE TASER STORY

O N A HOT AUGUST afternoon in 1993, I walked into the express elevator in the lobby of the John Hancock building on Chicago's North Michigan Avenue. I pressed the button for the ninety-sixth floor: the Signature Lounge bar. I grabbed a stool, ordered a Jack Daniel's, and sipped it while looking down on the city lights.

I had just turned twenty-three. I was about to receive my MBA from the University of Chicago's Booth School of Business and, like the other 580 graduates who would don cap and gown with me, I needed to decide whether to register for the school's job placement program. The corporate interviews on offer included some of the most prestigious consulting firms in the world, technology start-ups from Boston to Silicon Valley, and executive positions at a dozen different Fortune 500 companies. The whole point of business school, in a way, was to take advantage of the doors that swung open as soon as you were finished with it.

Something about that felt wrong to me. I don't know precisely what it was, but I had become possessed of an idea that I knew was worth more than a six-figure payday or a fancy corner office. I had been spending a lot of time thinking about a topic that few of my classmates were contemplating: killing.

NEARLY ALL entrepreneurial stories have a roundabout beginning, a moment when the first sparks of interest in an idea emerge but no one quite knows what the end result will be. Think of Thomas Edison telling Henry Ford to keep working on his quadricycle, or two teenagers named Steve (Jobs and Wozniak) using an illegal blue box to make free phone calls anywhere in the world.

I had a moment like that, too, but unfortunately, my story was more tragic than those stories. For me, it was when I heard the news about Todd Bogers and Corey Holmes. Todd and Corey had been friends of mine, a few years ahead of me at Chaparral High School. On December 7, 1991, they were driving on Scottsdale Road in Arizona when they got into an argument with a guy in another car named Kevin Osborn. He was angry, and he followed them to the parking lot of a Hyatt Regency hotel. Todd and Corey got out of their car and walked toward Osborn, who pulled out a pistol and shot both of them dead.

Far from the stereotypical criminal, Osborn was a white-collar professional—a business consultant with a legal firearm that he carried for personal protection. Unfortunately, he found himself in a situation where tempers flared, the gun came out, and things spun wildly out of control, ruining three lives in a matter of seconds.

It would be more than four years before Osborn was tried and convicted of manslaughter. It took under twenty-four hours, though, for the news about Todd and Corey's deaths to reach me. I was twenty-one years old, just a few months into my first year of business school. I was shocked by the sheer waste, by the wrongness of it. How in the hell did something like this happen, in Scottsdale and, of all places, in a hotel parking lot? What possessed someone like Kevin Osborn to carry a pistol? Shooting metal projectiles at more than a thousand feet per second to rip and tear flesh, inflicting gruesome injury or death in the name of self-protection, or in this case, in the name of road rage? It wasn't just wrong; it was pointless. Maybe Osborn thought he needed that gun to defend himself. But because we think of self-defense as the power to send metal projectiles through flesh, that gun also gave Osborn the terrible power to make a mistake that could not be undone. Couldn't we aspire to something better than that?

The shooting took place shortly before I traveled to Europe on an academic exchange program. If you've lived in a foreign country for a period of time, you know how disorienting and enlightening it can feel to be air-dropped into an entirely new culture. Some people are especially attuned to differences in the food, the child-rearing practices, or the politics. But for me, still raw and recovering from the murder of my friends, I was most shocked by Europeans' different ways of thinking about guns and violence.

The Europeans I met and befriended were flabbergasted by American gun violence. They didn't understand it, and they'd ask me—the representative of all things American—basic questions that I didn't have good answers to, the most essential being: *Why can't Americans figure out how to stop shooting each other?*

At first, I pushed back. I thought they might have an exaggerated sense of the problem, which came from viewing our use of guns through the lens of media and popular culture. But over time, their perspective also gave me the chance to think about the problem of guns with fresh eyes. I returned to their question: Why *don't* we find a way to stop shooting each other? Or, closer to home: Why were Corey and Todd shot to death?

I remember getting into long conversations with my European friends about what could be done to stop gun violence. They had suggestions: "Rick, maybe Americans could think more creatively about anti-violence education and ad campaigns." "Rick, maybe you could go into the schools and start talking to kids when they are young." "Rick, maybe your government should pass more restrictive gun-control laws."

Because I was born and raised in the United States, I knew that Americans have a different history with guns than Europeans do. You can argue for or against our cultural relationship with guns until you are blue in the face, but it is unlikely to change minds. In many areas of our country, one generation teaching the next about gun use and ownership is a point of pride, not a cause for concern. I do not mean to dismiss educational and legislative initiatives, but we have not been successful to date in substantially reducing gun-related deaths with those approaches.

Around this time, my mother called me in Europe. She wanted advice about buying a gun. There had been a murder not far from our Arizona home, and she felt unsafe. I reminded her about Todd and Corey, and I asked her whether a gun was really the best option. After all, it could be taken away from her or discharge accidentally. "Have you thought about something safer?" I asked. "Like a TASER device?" At the time, even though it wasn't a widely used product, the TASER system was available to purchase for your own protection. I had heard about it offhand, and I was looking for my mom to buy anything other than a firearm to keep in her home.

A couple of days later she called me to tell me that she had purchased a .38 revolver and signed up for shooting lessons. The rest of the conversation went like this:

Me: "But what about the TASER stun gun?"

Mom: "Can't buy one. I asked at the gun store and they said they're illegal."

The original TASER stun guns used gunpowder to fire non-lethal darts. But because the devices looked more like a flashlight than a pistol, they were considered Title II weapons and required a special license to purchase. That designation put this non-lethal weapon in the same category as grenades, sawed-off shotguns, and rocket launchers.

I was incredulous. My mother could legally buy an assault rifle, a twelve-gauge shotgun, or any one of a hundred different hunting rifles or target pistols, all of them explicitly designed to kill. But she couldn't buy a TASER device, a weapon explicitly designed *not* to kill?

An early-twentieth-century historian of technology, A.P. Usher, wrote that innovations tend to follow a pattern in which the first two steps are almost inevitable: recognizing an unfulfilled need and then finding something contradictory or absent in existing attempts to meet the need. Usher called this kind of failed attempt an "incomplete pattern," because something in the process of filling the need is still missing. The incompleteness in the pattern around guns and self-defense was obvious enough: rational people don't want to kill anyone; they just want to keep from getting killed. A tool that could accomplish the second while avoiding the first would complete the pattern. Something like a TASER weapon.

When I suggested to my mother that a TASER device, a remote stun gun that fires electrified darts, might be a better choice than a handgun, I'm honestly not sure where the idea came from. By the 1980s, well before I ever worked in this field, "TASER" was a famous brand associated with dart-firing electric stun guns. Maybe I remembered seeing terrorists use one on the mayor of San Francisco in the Dirty Harry movie *The Enforcer*. Or Jean-Claude Van Damme dodging a TASER weapon aimed at his head in *Timecop*. Maybe I read a newspaper article about it.

After failing to talk my mother into buying a stun gun, though, I became a sponge for information about them. Since I could not buy one of the dart-firing TASER weapons, I bought the legal versions of any stun gun I could find, annoyed the hell out of classmates with engineering degrees, and sat up late examining the parts of my dissected stun guns using references in my high school physics textbook. (Early stun guns were distant relatives of TASER weapons. The main difference was that you had to make physical contact with the target to deliver the charge, because the weapon did not fire projectiles.)

A back-of-the-envelope calculation showed that, with properly designed circuitry, a few small batteries could carry enough charged particles to make for a very powerful weapon. That idea was the basis for a second-generation electric weapon: a wireless projectile that could have a built-in electric stun circuit.

When I returned to the university for my last semester, I took a class on entrepreneurship. For my final paper, I wrote a business plan for a company—to be named Thundervolt Industries—that would produce and sell a non-lethal bullet, inspired by the TASER weapon. I got an A on the paper, and I was still thinking about that idea that night at the top of the John Hancock building. As I set down my drink, I knew I wouldn't enlist in the job placement program. I had arrived at the Signature Lounge with one of Usher's incomplete patterns. I left knowing how I would try to complete it.

WHILE I was eagerly consuming information about everything related to non-lethal, electric weapons, I kept stumbling on the name John "Jack" H. Cover. It seemed like he was on every patent related to this subject. I'd look up a paper related to the topic, and almost inevitably, Cover's name would be on the list of authors. I was desperately trying to find research or an expert who could help confirm whether my idea for a wireless electrified projectile could work, but there wasn't enough information in the published literature. Finally, I decided it was time to go to the source and talk to him about a problem that seemed to obsess both of us.

In the 1960s, Jack Cover had a prestigious job working on NASA's Apollo moon landing project. Then he heard about a commission organized by President Lyndon Johnson's Law Enforcement and Administration of Justice task force that recommended that the law enforcement community find better and more effective non-lethal weapons for situations of public unrest. In 1967, that was a real concern: Vietnam War protesters on college campuses, race riots in the streets, and police departments with few effective tools to keep the peace.

Cover came to be obsessed with the idea of creating non-lethal weapons that were as effective as their lethal counterparts. Around the same time that he read about the presidential commission's challenge to develop better non-lethal weapons, he read about a man who had come in contact with a downed power line and, amazingly, survived.

The man had apparently grabbed hold of a fence to climb over it. Unbeknownst to him, the fence had connected with a downed power line some distance away. Fortunately, the physics of the situation allowed only enough current to flow through his body to paralyze his skeletal muscles, not to kill him. He was frozen to the fence, unable to move, unable to let go, but conscious and alive. A bystander managed to push the man off the fence, breaking the connection. The man was stunned and shaken, but otherwise healthy.

Bingo, Cover thought. What if electricity was the force that could stop someone but not harm or disfigure them? Then another piece clicked into place. Much like me, Cover grew up a fan of science fiction. While I had *Star Wars*, *Star Trek*, and *Battlestar Galactica* in the 1970s, Cover grew up in the 1930s when science fiction was delivered in print form. The hero of the day was a mythical young inventor named Tom Swift, who would find himself in situations where the world was at risk and he would have to invent ways to save it. One of Cover's favorite novels was *Tom Swift and His Electric Rifle*, about the invention of a magical new weapon that fired bursts of electricity at its target.

The president's challenge to the nation to develop non-lethal weapons came together in Cover's head with two other thoughts: electricity can paralyze people without leaving long-term harm, and Cover's boyhood hero Tom Swift had created an electric rifle. Cover walked out of his rocket scientist job at NASA and set out to create a real-life version of Tom Swift's Electric Rifle.

Working out of his garage, and drawing on his deep network of scientists and medical experts from the space program, Cover began to construct prototypes that became the forerunners to modern TASER technology. Perhaps even to his own amazement, the technology worked. Using tens of thousands of volts of electricity, he was able to stun volunteers to the ground, yet moments later, they arose safe and unharmed.

When it came time to name the device, Cover and his team settled on TASER, an homage to Cover's fictional boyhood hero: "Thomas A. Swift's Electric Rifle." Science fiction became science, though it would be many decades before the garage devices he built ended up in mass production and in wide use in police departments.

In 1993, Cover was seventy-three years old and I was twenty-three. After seeing Cover's name on so many articles and patent applications, I decided to cold-call him. I dialed 411 and asked for the number of John H. Cover of Tucson, Arizona. Moments later, a gruff voice answered, and I was talking with the man who had invented the TASER weapon.

To say that he was skeptical of me, my intentions, and my experience would be an understatement. Jack spoke with an infectious enthusiasm about his invention, but also with the frustration of someone who had spent twenty-five years on a technology that could change the world and yet felt that he had been hampered and betrayed by government bureaucracy, untrustworthy business partners, and a public that wasn't yet ready for his invention.

When Jack invited me down to meet him, I immediately accepted. The next morning, I was standing on his doorstep. I remember waiting there, wondering where this was all going to take me, thinking I was nuts for passing up so many other opportunities through business school. When the door opened, there stood a beaming septuagenarian with a twinkle in his eyes.

Behind the door, I saw a veritable museum of weapons technology laid out around his entryway, living room, and dining room. Jack took me on a tour through these artifacts, and we spent hours walking down memory lane, as he told me stories about his adventures and his inventions.

"This one we designed for the James Bond types," he said, holding up what appeared to be a pack of cigarettes. On closer inspection, you could see a cartridge with two small bores protruding from the Lucky Strike packaging, the only giveaway that this was actually a sophisticated weapon disguised as a pack of cigarettes.

"This one we made for animal control. We called it the buffalo gun," he said, holding up a large, boxy gray rifle. Jack was eager to show me videos he took of himself using the buffalo gun to do exactly what it sounded like—stun a buffalo. Jack went on to tell story after story of how they had created an amazing array of prototypes and tested them on everything from Jack and his son to Bubbles, a hippopotamus that had escaped from a Southern California animal park.

Finally, Jack held up a device that resembled an old handheld vacuum cleaner crossed with a flashlight. "This was the final design we selected for production. We decided to make it into a flashlight design, so pilots could keep it in the cockpit to ward off hijackers, and it would still have a practical use as a flashlight." You could see he was disappointed that the final product was a bit more mundane than the James Bond version.

Then his mood darkened as he told me the story of what had stopped the TASER technology from succeeding. At first it was meddling managers from his primary private investor. Then it was the federal government, which stepped in a few months after launch and declared his beloved TASER system an illegal firearm. Following extensive production delays and increasing financial losses, investors lost hope when the federal government declared the TASER device a Title II firearm— the same highly restricted category as machine guns, short-barreled rifles, and sawed-off shotguns; and destructive devices including grenades, mortars, rocket launchers, large projectiles, and other heavy ordnance. Needless to say, it requires special permits to buy things like mortars and rocket launchers; this ruling effectively meant the TASER systems would be illegal for most people to buy. Because the TASER devices they had sold were not properly marked as firearms with the required serial numbers, and because the purchasers had not gotten the appropriate firearm licenses, they were recalled from the market and destroyed. The primary investor lost faith, and the company went under. Jack was crushed.

Jack raised money to give it another try, relaunching the product in the 1980s. This time it was properly marked and licensed as a firearm. However, the second iteration of the TASER company was also doomed to failure. It established a small business selling a handful of devices to police forces (including the Los Angeles Police Department), but the sales were insufficient to support the business, and for a second time, "TASER" as a business went under.

I had a sense, even in our earliest meeting, that I could do what Jack had not been able to do. Jack had a brilliant, inventive, fertile mind— he was overflowing with ideas and enthusiasm. But he was focused on inventing, not on building a sustainable business. His garage was a mess of brilliant but discarded projects, and in each case, he stumbled when it came to figuring out how to take an idea in his head and turn it into a product people would pay for. Sadly, this brilliant man had spent a lifetime in financial difficulty, which was partly due to the fact that, as smart as he was, he lacked the right skills or partners to take his creations and make them profitable.

I wasn't a physicist and I had nothing more than a textbook understanding of how electricity could stun a human being, but I believed that together we could take Jack's ideas and build an operation around them—an infrastructure and a sales force. I had learned how to take the concept of non-lethal weaponry and explain it to people in clear, non-scientific terms. I also believed that I had the skills and ability to get other people excited about this vision.

Like many a great inventor before and since, Jack wasn't the easiest guy in the world to work with; he was extraordinarily intelligent, but sometimes prickly and short-tempered. If we wanted to take his rudimentary TASER devices and make them more than obscure gizmos, we would have to figure out how to present them to the mass market.

FROM THAT first interaction, we recognized our mutual obsession with this problem. It began a long and winding road that took us from the idea-in-a-garage in 1993 to an initial public offering on the NASDAQ in 2001.

Over the last twenty years, TASER weapons have prevented about 200,000 deaths and serious injuries. Over 95 percent of public safety departments in the United States use TASER devices, and they are deployed approximately a thousand times a day. In every state in the United States, as well as 107 other countries, you will find one of our company's devices in the hands of law enforcement officers.

I don't share that information to brag: I'm sharing it in the spirit of constructing a case study. I'm trying to illustrate, through concrete numbers, that widespread adoption of new—even "radical" or "controversial"—technologies is possible. Often, when we speak of "new technologies," what people have in their minds is the latest app or newest gadget. Our vantage point for discussions about technology is often limited to low-stakes consumer technology. But my life experience shows that even professional-grade technologies in fields that we'd prefer not to think about can be improved upon.

Those technologies can not only be made better and be more widely adopted, but they can also be adopted by people who tend to be suspicious of anything new. People in government bureaucracies, accustomed to cautious, incremental, risk-averse action and long periods of contracting, are often the least likely to embrace the latest-and-greatest gadget. Those were the purchasers TASER technology needed to win over—and I'm proud to say it did.

When Jack and I were first starting out, we had our share of skeptics. That included the most skeptical group of people the world over: police officers. Cops are accustomed to people lying to them all day, every day. "No sir, I wasn't speeding." "Those drugs? Not mine!" Because they constantly see the unsavory sides of human behavior, police officers are, by nature, suspicious. They don't want stories; they want facts. They don't trust tall tales; they trust evidence.

So my initial conversations with law enforcement about TASER technology usually resulted in one of the following responses from police officers:

"This is America, son. We don't go around electrocuting people."

"I've got my fists and my gun. What do I need that thing for?"

"It will be too confusing to have to decide whether to shoot a TASER device or my gun."

"Some people just need to be killed."

The idea that a cop would carry anything other than a firearm was laughable. I often think about that laughter when I pitch the idea that killing is a technology problem or that the bullet can be made obsolete. Because today, the reaction of police departments to non-lethal weapons couldn't be more different.

A quick detour for a story: It's April in Florida, and with the approaching summer, the night air is already moist and humid. A highway patrolman is cruising on the interstate when an urgent call comes over his car radio: there's a disturbance at a residence involving a woman who's belligerent, possibly intoxicated, and armed. The address is a five-minute drive away, so the officer radios back that he's en route. He puts on his sirens and speeds to the destination.

When he arrives, two other officers are already at the scene, a darkened, one-story house. The other officers are posted at either side of the screen door, their handguns drawn at their sides. The highway patrolman draws his handgun, edges up to a safe distance, and tries to communicate with the woman through the screen. From the radio dispatcher, the two officers already on the scene, and his communication with the woman, he's able to piece together the story: she's recently had two children removed from her care by the Florida Department of Children and Families, she's deeply distraught, and she's talking about hurting herself.

In the moment, the cop makes a judgment. He looks at the house, hears the grief in the woman's voice, and realizes that she isn't homicidal—she's suicidal. She is attempting what is known as suicide by cop. She would leave the police no choice but to shoot her. Sensing this, the patrolman holsters his handgun and reaches for his TASER instead.

Seconds later, the woman kicks open the screen door, brandishing butcher knives in each hand. The patrolman fires his TASER device, hitting the woman in the chest and rendering her immobile on the ground. He and the two other officers are able to remove the knives from her clenched hands and to handcuff her without resistance. As they walked her to a waiting police car, one of the officers hears her mumble, "I'm sorry."

Soon after, the woman's family members arrive on the scene. Seeing the police cars with their lights flashing and an ambulance that has been called to perform a medical evaluation, they think that the woman has been shot dead. In statements given to the police, they confirm that the woman had discussed her plans to provoke a police officer into shooting her. They aren't surprised that she has gone through with it; they are surprised that she is still alive.

The story has a postscript, and it takes place several years later. The patrolman who fired the TASER weapon is eating at a local restaurant, when he recognizes one of the servers: it is the woman whose attempt at suicide by cop had failed on that April night, because one of the responding officers was equipped with a non-lethal weapon. The woman recognizes the patrolman, too. She points him out to another employee and says, "See that guy? He saved my life."

Unlike the futuristic scene in Raqqa, this story of a patrolman who avoided suicide by cop is real. It happened and the police officer shared it with me. Suicide by cop (SBC) is a real phenomenon—and it illustrates just how perverse incentives and behaviors can become when police officers have the ability to take a life. The term goes back to the 1950s, and by one estimate, almost 10 percent of the police shootings that happen every year are attempts at suicide by cop. Dr. Laurence Miller, a clinical and police psychologist, notes that while some incidents evolve in the moment into suicide by cop shootings, many are planned: "While some SBC incidents arise spontaneously out of the anger and panic of these situations, a good number of them appear to be planned, as shown by the fact that in nearly a third of SBC cases investigators find a suicide note that apologizes to the police for deliberately drawing their fire."

It wasn't suicide by cop that led me into this field, but during my time in it, I've seen a shift from the police who laughed at us when we first suggested the idea of non-lethal weapons. Today, public safety officials and front-line officers are begging for better alternatives to guns. It's not because they've all "gone soft." These days, it's because many of them have seen their colleagues dealing with the difficult aftermath of a police officer killing a civilian.

Stories of police suspects being shot and killed flood the headlines; the proliferation of cell phones and the instantaneous reach of social media have turned every officer-involved shooting into a potential worldwide controversy. It goes without saying that those cases and their aftermath are horrific for the families of the killed and the communities in which they occur. But the idea of being at the center of another case like Michael Brown's, Philando Castile's, or Tamir Rice's is every officer's worst nightmare as well.

Those cases have placed a renewed focus on non-lethal weapons. While they can't cure the problems of imperfect information and racial bias that so often come to the surface in these shootings, they can make errors in judgment result in far less devastating consequences.

THE RAPID expansion of the use of TASER weapons was not without problems. The high degree of effectiveness and relatively low risk of injury led some officers to become over reliant on their TASER. In some cases, this meant officers were using excessive force by moving directly to a TASER device deployment when they should have been using better verbal skills to de-escalate. In other cases, officers were using a TASER weapon in lethal force encounters where it was also inappropriate from a safety perspective.

In light of these issues and growing concern about police overusing or even abusing TASER weapons, I asked my team to help develop more advanced oversight systems. This was the beginning of the body camera technology—a tool my company developed specifically to help mitigate the risks of TASER overuse, with the additional benefit of helping defend officers who used the technology appropriately. The company expanded its mission of "Protect Life" to include "Protect Truth."

Over the past decade, body cameras have become standard equipment in many public safety agencies. As its product portfolio expanded, TASER International changed its name to Axon Enterprise to encompass a broad suite of technologies, from non-lethal weapons to wearable cameras to cloud software and artificial intelligence.

Only in the past two years have I started to publicly share our goal of "making the bullet obsolete." I anticipated a backlash. Police officers have a close relationship with their firearms. Even though we've gotten TASER products into their hands, the firearm remains the device that they believe stands between them and potential death. I was concerned that voicing our long-term goal of making the bullet obsolete would be met with skepticism or even outrage. But I was pleasantly surprised that the reaction has been overwhelmingly positive. It's fairly common now that an officer or police chief will come up to me and ask, "How are you coming on making the bullet obsolete?"

In other words, a truly transformative technology can penetrate even the most tradition-bound, this-is-the-way-we've-always-done-it culture. And this story is set to repeat itself over and over again—in the military, in policing, in criminal justice, in self-protection. New technology follows a familiar script: first the technologies will be mocked, then they'll be tested by a few early adopters, then those adopters will become evangelists, and finally, the technologies that were once laughed at become standard issue.

5

VIOLENCE IS LIKE A VIRUS: PUBLIC SAFETY

VIOLENCE IS CONTAGIOUS. It travels through our social networks and webs of relationships. When those close to us are "infected," we're much more likely to be infected ourselves.

Think of the way a flu spreads in wintertime. A coworker sneezes into their hand, and then a few seconds later opens a conference room door. A minute later, you touch the same door handle, then rub your nose. Now the virus is incubating in you. You don't know you're sick yet, so you don't take extra precautions like washing your hands frequently or using hand sanitizer, so, next thing you know, you've passed the flu to your spouse and kids, as well. Back at the office, other people who touched the same door handle may be spreading the same flu to their families in the same way.

It's nearly impossible to single out one person who bears "responsibility" for an outbreak. It's much easier to imagine the virus traveling from node to node through a network of connections. Some nodes in the network may be especially important, that is, the center of dense hubs of connections to other nodes. Maybe your coworker is a serial sneezer who never washes their hands. But more realistically, most nodes are of roughly equal importance. And the closer you are to an infected node in the network, the more likely you are to get infected yourself.

Using sophisticated software and mathematical modeling, epidemiologists are able to track the spread of diseases in just this way. This is less critical for something like the common cold or flu, but in the case of a deadly Ebola outbreak, for instance, these tools can be crucial to predicting the future spread of the disease and to saving lives.

Remarkably, violence operates much like the flu I just described. Sure, some violence is truly random—the terrifying, but thankfully rare, cases of "the man in the bushes" or of serial killers who select their victims on a whim. But violence is far more likely to travel through social networks. Most of the time, victims and perpetrators of violent crimes know one another. Several studies have shown that, with respect to sexual assaults, the majority are perpetrated by someone known to the victim. They're part of the same networks. They share friends, families, neighborhoods, hangouts. Further, friends and family of victims are far more likely to be involved in further violence themselves, whether as victims or as perpetrators. In this way, violence propagates itself through networks, just like the flu does.

For a simple example, think of the famous Hatfields and McCoys. During their decades-long feud in the nineteenth century, a victim of violence in Pike County, Kentucky, was extremely unlikely to be random. Odds are, that victim was part of the Hatfield–McCoy network, people who knew each other, communicated with each other, and fought with each other. For a more complex case, think of the dozens of interlocking gangs on the South Side of Chicago. Again, both victims and perpetrators of violence are likely to be part of that network, and violence spreads across it, usually between connected nodes, in a fairly predictable way.

Andrew Papachristos, a professor of sociology at Northwestern University, found in one study that being a member of a certain social network increased the odds of being a victim of homicide by 900 percent. It should go without saying that thinking about violence as a contagion should do nothing to lessen our sympathies for victims. Victims may be bound up in networks of violence, and they may be friends or family of perpetrators of violence—but there's no rational world in which that should be a death sentence. Thinking about violence as a virus isn't just a metaphor. It tells us that killing rarely happens out of the blue; rather, it spreads in ways that we can anticipate and predict.

And that knowledge—combined with advanced technology—can be decisive for the future of violence prevention. Right now, law enforcement is more like firefighting than epidemiology. Law enforcement officers are charged with responding to, containing, and investigating violence that has already taken place. Patrol officers spend the majority of their time responding to 911 calls. Once the short-term situation is under control, detectives and technicians take over, collecting the evidence, following up with an investigation and interviews, and working with a prosecutor to bring a case to trial as appropriate. True, there are educational programs that seek to deter kids from taking drugs, and programs meant to deter violence, but the proportion of time and energy spent on prevention of violence is relatively small compared to the time and energy spent responding to crimes that have already occurred.

William Bratton, former commissioner of the New York Police Department (as well as superintendent in Boston and chief in Los Angeles) is one of the most influential law enforcement leaders of the past fifty years. He led a shift in policing to put greater emphasis on the prevention of violence and crime. In his words, "Police can and must stay focused not only on the prevention of crime, but also on preventing disorder."

Bratton implemented a system called CompStat at the NYPD, which was later replicated across the country. In the CompStat system, commanders are measured by key statistics for their geographical areas, such as violent crimes, property crimes, vehicle thefts, and so on. CompStat incentivized these commanders to reduce reported incidents of crime or improve key measures of community safety. The system is widely credited with driving down crime rates in New York City. However, civil liberties groups cried foul at the aggressive use of pre-emptive measures, such as stopping young men (usually men of color) for random questioning and searches. Bratton himself agreed: "CompStat was misused in the 21st century in a quest for numbers-driven, activity-driven policing . . . the skyrocketing rate of enforcement in the form of arrests, summonses, and the issue that metastasized so drastically in New York City: the 'stop-and-frisk' policy."

An additional challenge of shifting toward crime prevention is that the vast majority of the data is retrospective. CompStat reports contained historical data and trends, but most of the forward-looking prevention work relies upon the intuition of area commanders and officers. Intuition is one of the most powerful tools of any officer, but it is always susceptible to bias, whether racism or any other form of pre-judgment that relies on subconscious "gut feel."

If we are going to truly shift toward preventing violence, public safety officials need the powerful tools of epidemiologists to help them identify the key nodes and future hotspots of violence. They need to predict future violence and cut it off at the source.

LET'S THINK through some hypotheticals about one possible version of that transformation. Imagine that we're in a dense urban area—say, Los Angeles—a few years in the future. In this scenario, LA is experiencing a modest increase in the rate of violent crime. But the LAVPD—the Los Angeles Violence Prevention Department—knows that the most efficient solution doesn't lie in behaving like firefighters, racing to respond to incidents after they occur. So the department doesn't direct more resources to putting patrol cars on the street or even to solving crimes after they happen. Rather, it invests in studying neighborhood networks of violence, discovering patterns that emerge, and anticipating the flow of violence to different nodes in the network.

Here's an example of what that might look like. By studying public social media posts, the LAVPD knows that two young men in central LA—let's call them Adam and Bob—are members of rival gangs. The previous year, Adam was in an altercation with Bob's cousin and was charged with a misdemeanor. Bob hasn't committed a crime, but he's still on the department's radar as someone who may be motivated to seek revenge in the future. So the department's AI system monitors his public posts for violent keywords and cross-checks anyone tagged in them against its database of people recently charged with crimes.

Sure enough, Adam and Bob's feud heats up over the summer. Ugly words and threats are exchanged, and soon enough, Bob says he's going to find Adam and he's bringing his gun. The keywords set off alarm bells, and the LAVPD escalates the situation, applying for a tracking search warrant from the local courts. The warrant is granted swiftly as the court AI system reviews the parameters and recommends approval to a judge who can quickly ascertain the facts from the alert and approve the warrant from their smartphone. The warrant allows for the tracking of Bob's phone and computerized monitoring of his data communications. Under the terms of the warrant, only messages that are relevant to the potential threats against Adam are shown to a human analyst. The systems are designed to protect the privacy of the individual by using computer systems to monitor content and only release relevant messages related to the threat for a human operator to review. The first-pass review of each message is anonymized so that the human operator is unaware of the subject's identity. This allows a human operator to first verify the relevance of the message to the threat listed in the surveillance warrant. Once relevance is established, the full message context and identity of the sender is released to the on-call investigator.

With the tracking search warrant in place, the LAVPD system tracks the location of Bob's phone, and compares its position to the location of Adam's phone to monitor for a probable interaction. The tracking system intercepts a text message from Bob's phone indicating he's heading to a specific convenience store to kill Adam. The AI models indicate a 90 percent probability that a violent confrontation is imminent. The human investigator concurs and approves an alert to a dispatcher in the 911 call center and the two closest patrol vehicles with the location, photographs of Adam and Bob, and a warning that Bob is considered to be armed and has expressed an intention of armed assault.

LAVPD has access to a network of approximately thirty thousand cameras across the city, some of which are owned by the agency, but most of which are owned by private businesses that share their camera feeds with the agency in exchange for an improved sense of security and crime deterrence. Cameras across the city activate based on the locations of both Adam's and Bob's cell phones, and a license plate recognition system identifies Bob's car at multiple points as it progresses through the city.

Adam is hanging out on a corner outside a convenience store. The store participates in the camera access program, and there is a police camera pod thirty yards down the street. Both cameras are tracking to the location of Adam's cell phone, and given the severity of the risk assessment, facial recognition software is cleared for use. The cameras zoom in on Adam and report a 92 percent match probability, which, given phone location concurrence, is escalated to 99 percent.

Now, under our current model of law enforcement, the odds of stopping Bob's crime before it occurs are vanishingly small. For the most part, they hang on the odds of a patrol car or an officer on foot being within a few feet of the convenience store just as Bob arrives—and even then, it's unlikely that they'll react in time.

But in this scenario, the LAVPD is able to quickly and pre-emptively marshal resources. Rather than gambling on the tiny odds of being in the vicinity by chance, the agency rushes an unmarked vehicle to Adam's location ahead of Bob's arrival. Soon enough, Bob pulls into the parking lot, gets out of his car, and moves toward Adam, hand clenching a gun in his pocket. But as he approaches Adam, an LAVPD officer gets out of the unmarked car, steps forward, and engages with Bob. Startled, Bob begins to pull his hand out of his pocket in the motion of a weapon draw. Without hesitating to see what comes out of the pocket, the officer immobilizes Bob with a non-lethal weapon. Bob falls to the ground, gun clattering across the sidewalk next to him, and he is quickly restrained by the LAVPD officers on scene. The entire sequence of events is captured on cameras worn by the officers involved, the store's security camera, and the agency's nearby camera pod.

BEFORE WE GO any further, I'm well aware that there are any number of valid concerns about the system I'm describing. Violation of privacy and mass surveillance are two that immediately come to mind. I don't want to blow past those criticisms; our communities deserve leaders who take them seriously. The balance of public safety and privacy is an important one. We need to address the concerns that communities have about bias, risks in the system, and the use of data to drive over-aggressive policing. These concerns are real, and they deserve to be treated seriously. A carefully designed system of checks and balances is necessary. Processes like data anonymization and detailed logging of the identity of officials accessing sensitive data can help to mitigate risks of misuse.

I'd point out, though, that I'm not describing a system from science fiction. All of the data to create such a system exists today, and the dangers of its misuse do, too. The City of Chicago already deploys a network of over tens of thousands of security cameras, including 3,700 owned by the city and another 32,000 through partnerships. That's an overwhelming amount of information, more than can possibly be looked at by human eyes on any given day.

There are several companies today offering systems that are starting to use data analytics for "predictive policing." PredPol and HunchLab are two of the early leaders in this field, using historical crime data to predict patterns of where crimes are more likely to occur in the future. Their software then highlights hot spots where public safety agencies can deploy officers in attempts to dissuade criminal activity.

Predictive policing is an understandably controversial topic, with critics fairly pointing out the risks of biased algorithms that lead to over-policing in communities of color. These risks are real, and important academic and private research and development efforts are underway to combat those biases. However, data-driven analysis shouldn't be written off entirely because of those concerns. It has shown the potential to improve outcomes in virtually every field of human endeavor, and policing is no different.

Set aside, though, new frontiers in data mining or the handing over of vital personal information or even sophisticated predictive policing technology. Today, police departments know how much valuable crime-solving information can be gained from rudimentary social media posts. Consider Robert Bowers, who, prior to committing an October 2018 massacre at a Pittsburgh synagogue, had broadcast his anti-Semitic views all over social media.

The challenge is that public safety agencies did not have sufficient tools to make use of this information proactively, before the massacre. A violence prevention effort of this kind would require an evolution in mindset and resource allocation, one that makes better and more systematic use of data that is already available in people's public statements posted on social media, while recognizing that we will need robust and responsive oversight systems to ensure the right balance between privacy of the individual and safety of the public.

To truly make sense of the deluge of information across social media, law enforcement would have to invest in AI tools capable of correlating millions of already public posts for signs of trouble. Public safety agencies have already begun to use social media monitoring tools and other technologies that help them sift through the flow of public information. But they've hit roadblocks. Following a 2016 report from the ACLU that revealed that cities were monitoring target words such as "#BlackLivesMatter," Facebook and Instagram announced in 2017 that they were banning developers from using their data for surveillance.

All tools are capable of good and evil applications, and all tools have their benefits and costs. In the scenario I described, the LAVPD saved two lives. It saved Adam from likely injury or murder, and it saved Bob from a lifetime behind bars. Perhaps he'll be charged with possession of an illegal firearm or making violent threats—serious charges, but not life-ending ones. And perhaps by the time he's served a more limited jail term, the feud between Bob and Adam will have cooled. Or perhaps not; all we can do is to play the percentages in the best way available to us, weighing the likelihood of a life-saving outcome against a failed one and against the social costs. Is a scenario like the one I've described worthwhile? Given the alternatives, it may well be.

For every action, there can be unintended consequences. Expanding the amount of data and surveillance used by officials in the public sector carries risk. Around the world, governments have used and abused data to go after anti-establishment protesters, lock people up without trial, and push a particular political belief or ideology. History teaches us, time and again, that data given to governments must be carefully managed. It isn't that long ago that authorities in East Germany monitored the communications of their citizens—and stored and pored over that data, using it to police people's thinking as well as their actions. That history, but a few decades old, is a powerful reminder that we have to safeguard the data our government collects and monitor what it does with the data in its possession.

On the other side of this debate, however, are a powerful set of arguments as well. Ask yourself: Would it be worth monitoring the public social media feeds of high school students if it enables us to effectively identify high-risk students and prevent a tragedy like the 2018 murder of seventeen people at Marjory Stoneman Douglas High School in Parkland, Florida?

Much of the answer to that question will depend on how the information is used and what kinds of responses are allowed to the information we find. No one would agree that we ought to lock up a student simply because of an off-color remark made on a Twitter account, but by the same token, a persistent pattern of threatening posts or tweets ought to be taken seriously. In that situation, there's risk in taking action—but there's also risk in doing nothing. The question for us, and for the public officials whom we entrust to make these decisions, is how much and what kind of risk we're willing to bear.

The public doesn't have an absolute view about questions of privacy versus security. In fact, we've already decided in some cases that people's personal privacy is outweighed by public well-being. In the fight against child sex crimes, for example, an organization called Thorn sifts through publicly available information on the internet looking for victims and bad actors. This effort was spearheaded by actor Ashton Kutcher, and the goal is a noble one: to protect children from sexual predators and find those predators so they can be stopped.

Thorn's product, Spotlight, enables law enforcement to chase down leads and use data to combat sex trafficking and child prostitution. "Spotlight uses natural language processing and machine learning to help sift through hundreds of thousands of ads daily to surface those that may be children," Thorn CEO Julie Cordua told the media. And what's more, it's been successful: by Thorn's reckoning, more than eighteen thousand sex trafficking victims have been identified and more than two thousand traffickers stopped because of Spotlight technology and the law enforcement officials who use it.

Does this technology fall under the umbrella of surveillance? Absolutely. Spotlight's tech dives deep into the dark web, as well as working with partner platforms ranging from Facebook to Google, to find photos of sexually exploited children. It then connects the digital dots so that law enforcement officials can find the people who took the photos and who are exploiting these children.

Is this a program that would, if subject to a public poll, enjoy broad support? I suspect it would. It's hard to argue against a program begun by a well-known celebrity whose goal is the protection of vulnerable children. But now change the manner in which you learn about this program. Not by reading about it in this book or even answering a poll-ster's question. Imagine, instead, if you first heard this story through social media. The headline flashes in your Facebook newsfeed: "FBI Surveillance Tool Scans the Internet, Including Your Inbox. Your Pho-tos Are Being Watched."

The trust you placed in this program depends, to some extent, on how that program and its trade-offs are presented to you. And the dif-ference in your reaction to a program like Thorn and a program run by government agencies can be staggering. There's information that many of us would not want state or federal agencies to access that we are happy to allow an organization like Thorn to take advantage of. These debates are often less about values and concepts than they are about means and ends.

Which is why, I'd argue, in a world of imperfect outcomes and per-sistent trade-offs, the point isn't evaluating "surveillance" or "privacy" as one-size-fits-all concepts. You don't check one of the boxes and then ignore the other. You need to consider each potential step over those lines on its own merits, with its own benefits and drawbacks.

As I described earlier, we can find both creative inspiration and cautionary tales in science fiction, which illustrate both benefits and drawbacks in their most extreme sense. In the movie version of this scenario, *Minority Report*, the statistical results presented were stagger-ingly better (more fair, more just, and more accurate) than we've ever seen in the real world. Yet the clever plot twists lead to injustice, and the overall effect is to violate our sense of right and wrong.

We need to remember, though, that compelling on-screen stories require gripping conflict; without conflict, these stories would not be interesting. That's why, in fiction, AI leads to the Terminator, and the hopeful capabilities of pre-crime intelligence in *Minority Report* unravel in a singular tragedy so powerful that the public only remem-bers the tragedy. The phrase "Minority Report" has become shorthand for something frightening and Orwellian. Those are fictional illustra-tions of the risks, and it's important for all of us to remember that they are being used as narrative devices, not attempts to predict the most likely future.

I am not suggesting we ignore the very real risks of biased information, overpolicing, and the mistakes that technology can make. I am saying that understanding risks, costs, and drawbacks is the start of the conversation, not the end of it. If we threw up our hands and refused to act every time we identified a potential risk, it would be the end of progress.

CONSIDER AN example not from the worlds of policing or law enforcement. In the 1990s, the US federal government launched the Human Genome Project, a highly controversial, multi-billion-dollar effort to sequence and decode the full DNA of a human. There were many legitimate concerns about how this technology could be misused in all manner of ways: pre-emptive abortions based on arbitrary character traits, attempts to design a master race, discrimination against people based on their DNA. These were legitimate fears. In the United States, in the early twentieth century, the eugenics movement was alive and well, and sterilization that we would find abhorrent today was put into practice.

The Human Genome Project had to overcome these quite legitimate concerns in order to survive. One way these ethical and legal issues were mitigated was by using the Ethical, Legal and Social Implications (ELSI) research program formed by the National Institutes of Health. The program was designed to guide the research and help maximize the benefits while identifying the risks and taking steps to mitigate them. This process admitted that research carries risk, without allowing that risk to prevent research altogether. Risk requires thoughtful management but shouldn't immediately lead us to give up on science.

Today, the discoveries from sequencing the human genome are leading to cures for cancers and genetic disorders, and to other improvements in the human condition. The nightmare scenarios—clone armies or a twenty-first-century eugenics movement—have not come to pass, and the results of the research are driving some of the most important advances in medicine today.

As I began to consider the deployment of AI-enabled technologies at my company, we recognized a similar risk of misuse. We found inspiration in the ELSI model when considering technologies that carried possibilities for potential abuse. Before launching into product development using facial recognition or other technologies, which potentially invade people's rights and privacy expectations, we formed an AI and Policing Technology Ethics Board with prominent civil rights and privacy experts so that we could better understand the risks as well as the benefits. By doing so, we are also making an effort to be more transparent: to acknowledge that, yes, there are risks involved in sifting through terabytes of information, but that we can take steps to minimize those risks.

We were fortunate to have Barry Friedman of the New York University School of Law accept the invitation to join our advisory board. He is one of the leading authorities on constitutional law, policing, criminal procedure, and the federal courts. He is also the founder and director of the NYU Law's Policing Project, which is dedicated to strengthening policing through democratic governance.

Professor Friedman and the Policing Project have been strong proponents of using a cost-benefit analysis approach for evaluating new technologies and procedures in public safety. This approach looks beyond the simple financial costs of acquiring technology or the cost of hours of labor to implement a new technique. It attempts to quantify the social costs and benefits in the broadest sense possible, and it offers a useful way of navigating the complexities of new technologies in fields like policing, the military, and criminal justice, where the consequences can be serious enough that the costs and benefits of any new technology must be fairly and rigorously weighed.

Friedman writes of his methodology, "In this fraught space, there are three questions with which we should all be concerned. First, are the measures that we take to keep us safe efficacious? Second, even if they are efficacious, are the benefits we are achieving reasonable in light of the costs, particularly social costs, that are incurred? And finally, even if the benefits are worth it, are there distributive costs about which we should be concerned, which is to say, are the costs of policing falling inordinately or inappropriately on some segment of society?"

The cost-benefit analysis approach is a framework to compare new initiatives and technology to the status quo, both financially and socially. In a rigorous application of the model, evaluators attempt to convert the social benefits and costs into a specific dollar amount—which is, of course, no easy task. For example, Friedman points to extensive academic literature in both economics and criminology that attempt to quantify both the tangible costs of crime (e.g., medical bills, property damage, property values) and the intangible costs (e.g., pain and suffering, fear, loss of quality of life).

Friedman is more closely associated with the community of public safety oversight groups and activists than he is with traditional law enforcement figures. Often, I find myself on the opposite sides of issues with him, but it's precisely this give-and-take that has made his work on our AI advisory board so valuable. Advice from people who largely agree with you is far less valuable than advice from those who bring a different perspective. In addition to his views on effective police oversight, Friedman's cost-benefit analysis helps us approach changes in policing in a structured and rigorous way, not through the emotion and fear that tend to dominate the headlines.

Part of the reason his contributions have so benefited our efforts in the world of AI is because he's able to help us think through hypotheticals in which either the technology or the human operator makes an error. That's important: technologists and entrepreneurs are often so optimistic about the promise of new technologies that we neglect to consider the perils. Among other things, Friedman helps us doff our rose-colored glasses and think of cases where the technology doesn't work as planned.

Returning, then, to my LAVPD scenario: What if there were a failure in the system? Maybe the officers on the scene were uncertain about the suspect's physical appearance, or the facial recognition systems register an error. Maybe, in their haste to stop a crime, the officers arrived at a false positive—that is, they identified the wrong pedestrian and used their weapons on an innocent man. Maybe, in this scenario, Bob changed his mind and failed to show up, and instead, the officers used their weapons on Carl, who roughly matched Bob's height and build and who was hurrying into the store before catching the subway to work. Any one of those is a failure in the system, but none are terribly far-fetched.

I don't have to tell you how closely this resembles scenes that play out today across the country with painful regularity. But even in this scenario, there's a crucial difference between the future hypothetical scenario and today's reality. Perhaps the most important detail is that the officer chose to draw a non-lethal weapon, even knowing that the suspect was reportedly carrying a lethal gun. Today, the officer would have drawn and fired their gun, not a non-lethal weapon.

Under what condition would an officer draw a non-lethal weapon instead of a lethal weapon? *Only* if the officer knew that the non-lethal weapon would stop the threat more reliably and quickly than a bullet. That's not the case today. Almost every officer today carries some form of non-lethal weapon, whether it's pepper spray, a TASER weapon, or a baton. The critical difference is that these weapons are secondary weapons—they are not used in life-and-death situations. The gun is still the first choice. But let's say, for the sake of argument, that the current trends in technology continue and that non-lethal weapons replace firearms in short order.

Then consider the difference in our original case. Instead of an innocent man cut down in the street with a gun, we have an innocent man who has been harmed but has not been harmed in a lasting way. The innocent man gets to go home to his family. He gets to live to tell the story—for that matter, he can sue the LAVPD if he was targeted in error. He may be owed a settlement to make up for the pain he's suffered. But most importantly, we have wounds that still have time to heal; we don't have a life permanently erased. To get from here to there, we have to change the weapon in the officer's hand and the technology ecosystem that supports them.

HOW ELSE can technology from this hypothetical LAVPD case reduce violence? One element that changes the dynamic between the officers and the civilians is that the entire incident is recorded on body cameras worn by both officers.

This, of course, isn't some figment of a far-off future. Today, body cameras are already used by about 70 percent of the major police agencies in the United States. But I want you to imagine how the future scenario may improve upon even the technology of today. In this future scenario, a body camera could do more than just record the incident for evidence. The body camera, combined with sophisticated artificial intelligence, could generate a multimedia police report, saving officers huge amounts of time and paperwork.

Paperwork? You might be surprised that, in a discussion about AI technologies preventing crimes and weapons of the future preventing deaths, I'd raise the mundane topic of filing reports. And I admit, it's not the most exciting or talked-about issue—but it's a hugely important one for police officers and their departments. The automation of much of modern reporting for police officers could have a substantial impact on reducing violence.

That's because it would increase the time police officers are actually doing the work of policing, namely, interacting with the citizens they serve and patrolling the streets. In one of the most extensive time studies on police work, John Webster from the University of Illinois, Chicago, tracked the activities of the patrol division of an agency with nine hundred officers. He analyzed more than 300,000 hours of activity. And what he found was shocking: over 50 percent of officers' time was spent on administrative tasks—not what anyone would think of as ordinary police work. That study was conducted in 1970. Since then, the number of mandated reports has increased significantly; as new issues in policing have cropped up, departments issue new reporting requirements. For instance, many new mandated reports are intended to increase oversight around issues such as potential racial bias in police stops.

Automating reports from police videos has the potential to effectively double the number of officers who are out performing public safety activities. Such a change would make a difference for violence prevention. Why? Because research has shown that among the most effective measures for driving down violent crimes is to put more officers on the streets. And when it comes to putting officers on the streets, improving department efficiency and freeing up officer time is far more cost-effective than hiring more individual officers. This is especially true in our era of tighter state and local budgets, when many municipalities struggle to even keep staffing at current levels.

The time that could be returned to police officers is meaningful. Right now, an officer's day is generally split between filing paperwork and responding to the urgent crises that come over the radio. Regarding the kinds of activities that might help them build bridges with their communities—visiting schools, holding roundtables with community leaders, checking in with people they know are at risk—I often hear the same thing from police officers: "Rick, there just aren't enough hours in the day for that sort of thing."

In a survey of twelve thousand police chiefs and command staff, Nuance Communications found the following: "39 percent of respondents said they spent 3-4 hours daily on incident reports and other police paperwork. More than 30 percent of those surveyed said they spend at least a quarter of their day back at the station working on reports, not out on the streets or responding to citizens. And nearly 15 percent said they spent more than half of their workday on reporting duties."

What if there were more effective patrol hours per day and fewer hours of paperwork? What if, in a not-too-distant future, reports were largely automated by powerful algorithms that could look at a routine traffic stop and fill in a form, one that today might take an officer an extra hour? What if, in the future, that form were to appear prepopulated on an officer's cell phone or laptop so that they simply had to examine it for errors rather than manually completing it field by field?

The combination of body camera technology and advancements in artificial intelligence will make this possible. In fact, what I'm describing is simply an advanced version of the technology that has benefited consumers over the last ten years. Think of the relief you feel when your credit card information is already filled out on a website where you shop, or when your mailing address is already plugged in while you're buying airline tickets, or your phone number doesn't need to be typed in for the umpteenth time. Advanced and secure versions of that seemingly simple technology can be used to help police officers. I believe that change could put more officers back into the neighborhoods and communities they serve.

IT ISN'T JUST about making police work more efficient; it's also about making it fairer and more equitable. While concerns about AI technology focus on its misapplication, a wearable camera with accurate, real-time facial recognition might help officers avoid problems of misidentification—the exact kinds of problems that, today, lead to tragedies and accusations of racism against police.

If deployed with the right safeguards, those technologies might help defend against human bias. Let me be clear: I don't believe we have technology advanced enough today to be broadly applied, and neither the social costs nor the potential benefits of a broader application of AI-driven facial recognition have been adequately explored.

Today, facial recognition is not sufficiently accurate or reliable for real-time use in high-pressure situations. Algorithms are beset by bias problems, partly because the algorithms are based on historical data, which can often contain institutional bias. If, for example, there is a police agency with a systemic biased pattern in how it makes its decisions in the past, that bias could too easily be baked into the algorithms that predict criminality or wrongdoing. We haven't seen algorithms that I would trust to do more advanced police work correctly—yet.

If history has us shown anything, it's that technology moves forward, often more rapidly than any of us would expect. The boxy, early mobile phones were a pale shadow of today's most advanced smartphones, which are the equivalent of yesterday's room-sized supercomputers and now reside in our pockets. My smart watch contains more computing horsepower on my wrist than the spacecraft that first sent astronauts to the moon.

I believe bias and accuracy problems in artificial intelligence technologies are not only solvable, but can and must be solved quickly. It is unacceptable, for example, to ignore the value that facial recognition could bring to reducing societal crime, but it is equally unacceptable to use facial recognition this way until the bias and accuracy problems have been solved. More than the specific challenges of one technology, the broader point is that we need to separate the potential of today's technology from future use cases where that potential will be different. Our present limits shouldn't limit our future.

In the short run, society should decline to allow the use of technologies like real-time facial recognition, where the costs of the inaccuracies and bias outweigh the benefits we could expect. But I don't think that these concerns ought to impede our efforts to improve the technologies and to improve the cost-to-benefit ratio. Once a technology does prove itself to be efficacious and with benefits that outweigh the social costs, then it should be deployed. Over the longer term, it makes sense to probe what use cases should remain off limits, even if the limitations of the technology are overcome. That is to say, even if the technology worked perfectly and without bias, would we want to use it? And if so, in which use cases? And then, how can the right guardrails be created to limit its usage to those positive use cases and avoid the most damaging negative misuses?

Those aren't easy questions to answer, but they are important questions regarding everything from facial recognition technologies to predictive policing and advanced analytics. We need to chart a course from today's imperfections, navigating bias and inaccuracy, setting guide posts where the costs and benefits ought to prevent public acceptability of a technology. It's a great deal of work, solving problems with so much complexity and so many layers. But as I've said from the beginning, the status quo isn't perfect either: we have archaic systems beset with human bias. The goal should be better—not perfect.

ANOTHER TECHNOLOGY that contains both huge promise and significant potential for abuse: the use of armed drones and robots by police. This is no longer an abstract question or an idea from a sci-fi movie. In 2016, Dallas police officers used a robot rigged with explosives to kill a sniper who had taken the lives of five officers and was holed up inside a building. It was the first time a police robot had been used in this way in the United States, and it generated surprisingly little controversy— largely due to the dangerous and murderous nature of the target.

The robot was under police control, but it wasn't conducting a precision strike. The robot had a bomb attached to it, and the operation was a robotic suicide bombing mission to end the life of the suspect. While the robot could open a door, roam around a building, and remotely perform a variety of tasks, the only thing it could do when it engaged a suspect was blow itself up. It had no ability to control or interact with a suspect beyond hoping that the bomb on board would eliminate the suspect. As one American Civil Liberties Union leader, Scott

Greenwood, put it to me in describing this case, "The only reason he had to be killed was because they didn't have any better option."

I understand why the police made the decision to send in a kamikaze robot, even with the risks involved. The building had been evacuated and the officers anticipated, correctly, that targeting the sniper with a robot wouldn't cause any collateral damage. Tactically, it was the best available option: this was a lethal killer who had already taken the lives of five police officers. They didn't want there to be a sixth. In that moment, there wasn't any risk as great as the risk of another officer being killed. So the robot with a mission to kill was the safest choice of the available option set.

Part of the reason, I believe, that strapping a bomb on the robot seems acceptable to many is that we tend to tolerate extraordinary means during a crisis. Under normal circumstances, the effort to arm a robot with any weapon, including a non-lethal weapon, would generate big, salacious headlines—and those headlines would, I suspect, criticize the law enforcement agency in question for overstepping.

So it's politically risky to build a robotic capability until the moment when the threat is off the charts, at which point, there's not much time to engineer a sophisticated solution. Duct tape and some explosives is about the only option available on the fly—but it's also a politically and publicly acceptable option given the circumstances.

"If that robot had more dexterity, if it was faster, if it had more sophisticated weapons... he could have been incapacitated long enough for officers to take him into custody," Greenwood said to me. "This is hailed as a great innovative success in policing, but it's a technological failure." I'm hopeful that future robots of this kind will have the ability to stun or incapacitate suspects, for the reasons I've mentioned. Subduing a suspect—even one who has killed five police officers—and understanding why they did what they did, then using that information to prevent future violence, is more valuable than just neutralizing the threat by killing the suspect. The world is a better place, I would argue, if we can capture even the worst perpetrators alive.

When the topic of police drones or robots comes up, people tend to think of the nightmare scenario of armed drones patrolling our neighborhoods from the sky, dropping down to enforce the lightest infraction.

This use (and many others) is clearly unacceptable. However, if we think about the scenarios where police officers turn to assault weapons and sniper rifles today, such as the mass shootings at the Pulse Night Club in Orlando in 2016 or at the Mandalay Bay resort in Las Vegas in 2017, we can envision situations in which non-lethally armed drones or robots could save many lives and offer a far safer alternative than sending in SWAT officers armed with automatic assault weapons.

What I would hope comes out of the Dallas example is neither blind acceptance of a future of armed robots nor the belief that this is a one-off case that will never come up again. If we are willing to accept this kind of technology built in the heat of the moment, we ought to be able to have a sober and rational discussion about it in the light of the day. What should robots be allowed to do, and not allowed to do? Should they be allowed to kill, or merely incapacitate? How do we deal with a robot that misidentifies a suspect, and how much control should a human operator have over its actions? How does artificial intelligence technology affect what these robots can do?

These are important and urgent questions, and they aren't going to be given a fair hearing when there's an armed killer who needs to be dealt with. I do believe there's promise in these kinds of technologies to keep both police officers and the public safe—and that many of the alternatives today (e.g., police officers having to race at the suspect and risk their lives) are worse. But I also believe we need to have a rigorous public debate about these technologies and what safeguards we put in place to make sure that they are used appropriately.

I STARTED this chapter with a scenario in which violence prevention officers use their knowledge of networks of violence to predict crime and stop it before it starts by using non-lethal weapons. But ideally, I'd like to imagine a world in which the use of even those non-lethal weapons is rare. The same tools that pinpoint central nodes in networks of violence can tell us where to direct social services, conflict mediation, community peacemakers, and job opportunities. The point of mapping violence as accurately as epidemiologists map disease isn't simply to intervene with a non-lethal weapon at the last possible moment. It is, wherever possible, to prevent and deter violence at its source.

EXTENDING
THE TREND: MILITARY

HROUGHOUT MOST of history, the trajectory of weapons has been in the direction of ever-increasing range and lethality, from the handheld spear to the atlatl to artillery, battleship, aerial bomber, and the intercontinental ballistic nuclear missile.

Practically, the ability to inflict massive casualties hit a logical endpoint with the development of the thermonuclear weapon. While the fission-based nuclear weapons dropped on Hiroshima and Nagasaki represent the most devastating weapons used in actual warfare, the development of ever more deadly weapons continued for one more generation. The explosive power of the bombs dropped at the end of World War II was approximately fifteen to twenty kilotons, meaning the explosive power of these devices was the equivalent of fifteen to twenty thousand tons of TNT explosives. A single detonation in Hiroshima killed approximately 140,000 people, or 39 percent of the city's population.

And yet, in spite of that destructive force, both the United States and the Soviet Union set to work on an even more powerful and deadly device: the thermonuclear weapon. By adding a hydrogen fusion reaction to a nuclear weapon, they were able to increase the destructive power by another three orders of magnitude, or by more than a thousand-fold. The RDS-220 Hydrogen Bomb (known as the Tsar Bomba, or king of bombs) deployed over the Arctic Ocean on October 30, 1961, released fifty megatons of explosive energy, or the equivalent of the Hiroshima bomb *3,300 times over*.

The destructive power of a nuclear weapon was said to be the end of the road. However, the destructive power of a thermonuclear weapon at thousands of times greater surely is. This is borne out by the fact that there has been no effort to deploy or test a more powerful weapon in over fifty years.

Figure 1 illustrates this point. The fifteen kilotons of Hiroshima is the tiny blip in the lower left corner. This image surprises even military leaders familiar with bombs and their power.

What does this mean in practice? We have hit the end of the logic of weapons that kill on a greater and greater scale. In fact, we have made weapons impractically deadly now—to the point where warfare between nuclear powers is untenable, unthinkable, world-ending. That potential devastation is, perhaps somewhat ironically, arguably

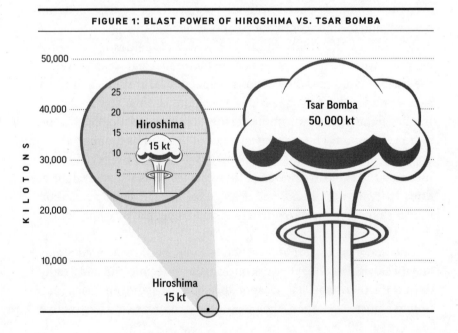

FIGURE 1: BLAST POWER OF HIROSHIMA VS. TSAR BOMBA

responsible for the more recent decrease in overall deaths related to war, what scholars call the Long Peace, covering the period after the end of World War II.

That is a historical anomaly. Throughout most of history, war was a logical choice for a king or emperor to exercise power. While the cost of war on a society could be significant, its impact on the leadership of the country making the decision could be neutral to positive. The human suffering of war occurred hundreds of miles away. And if an aggressor was correct in their estimate that the war was winnable, the empire or regime could emerge stronger, with control of greater territory and resources.

Nuclear weapons changed that calculus. Their sheer destructive force makes "winning" impossible against an adversary who is similarly armed. And even if a nuclear war could be won, the genocide it would inflict should be beyond the pale for even the most aggressive autocrat. All parties would be destroyed, an idea captured in the well-known phrase "mutually assured destruction," which prevented the Cold War from turning hot. Even as they rattled sabers, both the United States and the Soviet Union knew that if the conflict turned hot, it might bring about the end of modern life as we know it.

Thus, since Tsar Bomba in 1961, we have seen a fundamental shift in the course of weapons development, away from increasing lethality to increasing precision. Laser-guided bombs gave way to GPS-guided cruise missiles and ultimately to remotely operated drones that could loiter for extended periods of time, waiting for the most opportune moment to drop a Hellfire missile on a carefully targeted individual or group of individuals.

Typically, these missions would focus extraordinary effort on gaining sufficient intelligence and waiting for extended time periods for an opportunity to take out the subject with no or minimal collateral damage. In many cases, the final decision to deploy targeted lethal force would escalate up to the president or the president's immediate staff.

Never before in history had there been such a high level of care in making killing decisions or, perhaps more importantly, making the effort to carefully contain killing. The next logical step is to build even more highly targeted weapons. A concept video available online demonstrates one vision of the high-precision warfare of the future. It depicts a small drone that, if built, could fly up to a single individual, right next to their head, and then detonate a focused lethal explosion (see www.EndOfKilling.com/drones). The video makes a point about how such technology could be misused. However, if you stop the video at the three-minute mark, even before it dives into the misuse, there is still something about the footage that feels wrong. Even if the technology were carefully controlled, its intended effect still *feels* sinister. This level of lethal precision is unsettling.

Why are we put off by something like this? Why is it that a drone that flies up to someone's face and detonates a precision explosive makes us queasier than footage of a bomb that indiscriminately destroys a car and all of its inhabitants, which may house a terrorist but also their children? On the level of moral mathematics, the idea of a microdrone that is smart enough to follow the car, wait for the inhabitants to exit, and then use facial identification to ID only the terrorist and precisely and cleanly kill them seems like an objectively better outcome. But somewhere deep down, our subconscious senses something wrong about this.

Here's one reason: if we have technology with such advanced precision targeting, then killing seems anachronistic, even barbaric. Our subconscious hints at something morally wrong about remote precision killing, because it reveals the imprecision of our justice. After all, if we can get this good at taking a single life, shouldn't we be just as good at avoiding the need to take a life altogether?

This video challenges all of our usual justifications for sanctioned killing. We can't throw up our hands and declare, "I had no choice." Drones can't claim even a scintilla of self-defense as a motivation. (There is an argument that killing the terrorist is an extended form of societal self-defense. But, I would argue, the broader societal defense could be effectively achieved with a capture scenario.)

Surely, if we can create the technology of this lethal microdrone, we can create any number of choices that are preferable to killing. That's why arguments about these hyper-precision lethal weapons strike me as just as hollow as arguments in favor of bigger nuclear warheads. Much like the advancement of thermonuclear weapons reaches an absurd end state, precision killing does the same. Past a certain point, killing precisely seems no more right than killing indiscriminately.

I would propose a new focus, a marked shift away from both lethality and precision: reversibility. Precision has historically been about confining irreversible lethal force to a specific location. I believe the next frontier is to focus on precisely delivering reversible non-lethal force. That gives us a precision across both time and space. We can deliver the effect in a precise location, and if we get it wrong, the effect is not permanent so we can wind back the clock.

Why does reversibility matter? Because the information we use to go after anyone—a military target, a murder suspect—will always be imperfect. And if that information is wrong, we need to be able to undo any action we have taken. The move from increasing lethality to increasing precision has, no doubt, saved many lives and reduced the carnage from conflict during the Long Peace. But there is still a great deal of work to be done. The next logical evolution in advancing our weaponry seems to be a shift from precise lethality to reversibility.

THAT'S PART of why the warfare scenes in Raqqa that open this book are so compelling: we know that none of the decisions the drones in that scenario make are permanent. Every single one can be reversed if needed, the clock turned back to before the drone fired the dart or took the suspect up to the Kraken.

As I mentioned, that scene illustrates technologies that are either available today or are likely to be available within the next few decades, even if the application of those technologies is currently limited to the civilian context. Small flying drones with highly accurate flight control systems? You can order one from Amazon right now. Tiny electrical weapons capable of highly accurate targeting? All of the core technology exists today. Targeting algorithms and scanners that recognize faces, eyes, even fingerprints? Deployed today, in different forms and venues.

Progress in technology isn't just about the availability of the proper circuits and software. As often as not, changes in technology depend on people's willingness to change habits, not hardware—their willingness to take a risk on something that might or might not work in order to bring about something that works better. And in the case of the Raqqa scene, the habits that need to change are those of one of the biggest, most complex, most hidebound bureaucracies: national militaries.

We can imagine any manner of objections to the Raqqa scene. But as with many of the ideas in this book, it's a case that shouldn't be examined in isolation; it has to be compared to today's conflicts and warfare. You have to judge a future military action in Raqqa against the same scene unfolding today in the Syrian civil war or the war in Iraq—conflicts that have claimed untold thousands of Syrian, American, and Iraqi lives.

That's not to minimize the concerns about the future, or to simply decide that the future will be better than the present so we should charge headlong into it. I understand why people fret about a future with flying armed drones; as someone intimately familiar with the technology, I share their concerns. But I also want to make sure those concerns don't kill off good, potentially life-saving ideas before we've fully explored how they might work. Technology can inspire fear, but a *Terminator* nightmare scenario—in which armed robots run amok—is not the only possible vision of the future.

What about the concern, voiced by leaders like Elon Musk, that AI technology will one day escape human control and pose a risk far greater than the problems it purported to solve? I agree with him, broadly. We have to build in safeguards that prevent machines from making autonomous decisions without oversight. He's right in arguing that artificial intelligence is also the kind of genie that can't be put back in the bottle, and that any regulatory questions ought to be thoughtfully considered now, while we still have the ability to build safeguards. In the case of AI, an ounce of prevention really is worth a pound of cure— because we may not be able to administer the cure later.

This goes double for military applications of this technology. It's one thing for a Tesla to steer us to the wrong restaurant by accident, but it's entirely another if an armed drone makes an error. That's why I'm as staunch a proponent as it gets of human operators retaining the decision to use force, especially lethal force. And that's why I believe that drones and the surveillance technology that powers them should not be enabled to autonomously act to kill a civilian or soldier. In fact, we should avoid enabling those technologies to kill at all.

That said, in discussing the future application of these technologies, I think it's important not to become bound by how they operate today. When we focus on how these tools function now, the discussion about them too quickly goes from complex and nuanced to binary and reductive: you're either for entirely autonomous devices that can take lives at random, or you're against using remotely operated systems altogether. The debate, as it too often does, devolves into simple and simplistic caricatures. Either you're with the robots, or you're against them.

Consider these technologies more carefully and with an eye to what they might be able to do as technology becomes more advanced. These systems could be designed to de-escalate and de-risk situations. They could improve human judgment (for example, identifying the right suspect) or they might provide a safeguard if that judgment goes wrong (subduing a suspect as opposed to killing them). They can help us put a neutral device in a middle of a situation (such as a potential beheading) as opposed to sending in a platoon. Context is king.

It's changing a context that is often technology's most vital role. I tend to think of technologies in situations like the Raqqa scene as referees on a football field. They stop the action when a transgression has occurred, and they make a determination about who did what. And just as referees aren't at risk of being suddenly attacked by the players on the field, neither are semi-autonomous systems in a position where they need excessive firepower to defend themselves. They aren't combatants; they're arbiters, controlled by trained human analysts. And they can help us deal with high-stakes situations by lowering the risk of death, dismemberment, or permanent harm; thus, they could operate more systematically and humanely.

LET'S DEAL with the second-most-popular objection to futuristic weapons of war: surely the cost of a fleet of non-lethal microdrones would be prohibitive, right?

This objection needs to be met squarely. I doubt that anything we can put into the field would be more expensive than the way our wars are fought today. That's the simple and, in some ways, simplistic response. Here is a more relevant one for our purposes: a technology-first approach to warfare benefits from the same forces that have lowered the costs of our laptops, phones, and cameras for the last few decades, all while improving their quality.

That idea is known as the Law of Accelerating Returns, attributed to inventor Ray Kurzweil. It goes like this: computing technology is growing more powerful at an exponential rate, doubling in performance (for the same cost) about every one to two years. Many people know this concept as Moore's Law, named after Intel chairman Gordon Moore, who noticed that the number of transistors that could fit in a microprocessor was growing on this exponential curve. Another way of putting it is that electronics continue to grow exponentially cheaper, because over time, you can fit an exponentially larger amount of computing power in the same space. Moore's Law is a key reason why much more than the computing power it took to put a man on the moon in 1969 now fits in an Apple Watch on your wrist.

Those forces are already at work on the technologies I'm describing. Right now, the cost of even the most advanced drones is plummeting. And if drones are anything like any other new piece of technology, their cost will continue to fall. A hundred thousand drones that cost $2,000 each add up to a total price tag of $200 million—that's less than the cost of a single F-22 fighter. It's a fraction of the cost of a single nuclear submarine. Consider the Raqqa scene again: for the cost of one fighter jet, we could pay to cover the city with Dragonfly technology.

What about human transport drones? We already have early examples of drone-style flying cars being deployed in Dubai, and they soon will be by the likes of Uber, Boeing, and Airbus in the United States. The head of product development at Uber recently predicted they would be able to fly a passenger from San Jose to San Francisco—nearly a two-hour drive under normal conditions—in under fifteen minutes within the next five years. It does not take a leap of imagination to envision a winch and harness system integrated into one of these drones, combined with a robotic system to attach it to an unwilling subject, especially a subject who has been incapacitated by a conducted electric weapon similar to the ones carried by hundreds of thousands of police officers today.

What about neural scanning technology? This is probably the most far out of the technologies described in the Raqqa scene, closer to science fiction than science today. But it is closer than most people think. Today, a company called Openwater is exploring the frontiers of wearable technology using novel opto-electronics that are capable of scanning both structures and metabolic activity happening within the body. In other words, Openwater is developing a technology using wavelengths of light that can transmit into the body and measure the returning signals to monitor both physical structures and metabolic activity in real time. This technology could be used to find cancers or cardiovascular disease or, according to the company's website, for "communication via thought."

Present-day technology in university and corporate research labs already enables people by thought alone to answer simple questions, manipulate mechanical devices, or even play video games. It's reasonable to imagine advanced AI brought to bear on the results of these "brain scans," enabling scientists to develop an ever-expanding laboratory of correlations between brain scan data and subjective experiences or thought. In fact, Ray Kurzweil has predicted that by the 2030s such technology will make possible sophisticated "brain-computer" interfaces. Even Elon Musk has started a company called Neuralink, which he believes, according to a 2018 statement, will achieve high-bandwidth brain-to-computer communications in "about a decade."

The timelines put forth by Musk and Kurzweil are controversial. Many technologists point out that we do not yet even understand how the brain stores information, much less have the capacity to scan the trillions of synaptic connections in the human brain non-invasively. Regardless of whether the technology to scan and interpret information from the brain is twenty years or a hundred years away, the technology will be so transformative—and so incredibly invasive of personal privacy—that I believe the prudent approach is to open discussions today to mitigate the risks before they emerge. It would be better to be intellectually prepared if the technology evolves faster than expected than to leave policy completely unprepared until the technology is unleashed.

Governments are also investing in the field of brain research. The European Union recently launched the Human Brain Project, whose purpose is to map and fully understand the human brain, including how it processes and stores information. And in 2013, the White House launched something similar in the BRAIN (Brain Research through Advancing Innovative Neurotechnologies) Initiative, whose goal is to support the "development and application of innovative technologies that can create a dynamic understanding of brain function." Many people are skeptical of this project, but think back to the earlier example of the mapping of the human genome. It was greeted with the same skepticism. In the 1990s, when the US government launched the Human Genome Project, critics declared it to be a fantasy and a waste. Using the technology of the day, experts predicted it would take a thousand years to decode the DNA of a single human being, at an incalculably high cost. Why should taxpayers be on the hook for something like that?

They were right—sort of. Using the technology of the early 1990s, it would have taken over a thousand years and cost taxpayers trillions. What they missed was that the underlying technology was improving at an exponential rate. The computers used to process the data from gene sequencing machines were doubling in performance every eighteen months. This was Moore's Law in effect, and it applied to both the data processing and the underlying technology for gene sequencing.

At the outset, it looked like the project's critics might be right. Seven years into the Human Genome Project, the government had spent hundreds of millions of dollars and only 1 percent of the project had been completed. A thousand years was looking like the right timeframe for completion. Something to know about exponential technologies, though: they appear to be moving very slowly, then they explode.

That's not faith; that's math. Think about this number sequence, which is doubling on each line:

0.02%
0.04%
0.08%
0.16%
0.32%
0.64%
1.28%

If these numbers represented percentage of progress for each of the first seven years of the Human Genome Project, progress would have looked incredibly slow and would have been right around 1 percent complete. But then let's look at what would happen in the next seven years:

2.56%

5.12%

10.24%

20.48%

40.96%

81.92%

163.84%

Notice that it took us seven years just to get to 1.28 percent. Then, in the next seven years, we get to almost 164 percent—so the project is well past 100 percent completed. The progress in the first seven years is less than 1 percent of the final number.

And that is exactly what happened with the Human Genome Project. About seven years after hitting the 1 percent completion mark, the project was completed in 2003, at a total cost of three billion dollars. This exponential progress has continued to this day. A decade and a half later, you can go to a physician's office (or just mail in a swab sample from inside your mouth) and have your full genome sequenced in less than a day, for under five hundred dollars. Think about that. For less than five hundred dollars, you can achieve something that was thought to be impossible twenty-five years ago and cost three billion dollars when it first became possible fifteen years ago.

Had you predicted this in 1993, many of the world's experts on the human genome would have laughed. It is exactly this pattern that is now playing out with the Human Brain Project and similar research. Extrapolating from the exponential growth of computing technology, if the computing power available to map the human brain continues to double about every eighteen months, it would mean that the computers used to work on this problem in the 2040s will be about one billion times more powerful than the computers that cracked the human genome in 2003. And the resolution of the scanning technology to observe the activity and structures in a living brain are also doubling in capability at a similar rate.

So if researchers today can put a cap with electrodes on your head and observe real-time brain activity tied to specific cognitive functions, it seems possible that we could achieve high-resolution brain scans in another twenty to forty years. Or at a minimum, highly reliable lie detector technology (which should be orders of magnitude easier than full-resolution information mapping of the brain).

Even if you set aside those specific applications, the point is this: two decades is an eternity in technology terms, especially when we know that the technology we are building has an accelerating rate of improvement. Over the course of that time, the technology that today is considered expensive, wasteful, or potentially dangerous could come to be as acceptable and natural as swabbing your cheek and analyzing your genes—a service once thought prohibitively expensive and impossibly complex that can now be delivered to your door by consumer services like 23andMe for under a hundred dollars.

BY NOW, hopefully you accept that it is at least possible that technologies for futuristic warfighting could be developed in the coming decades. The next question is whether we *should* develop them. One of the prime objections to developing these technologies is, ironically, the resistance of some people to having the military use them. But just because I've outlined scenarios in which the military uses a given technology, that doesn't mean the military will be the one developing it—nor does it have to be specifically designed with the goal of apprehending terrorists or keeping the peace.

That's already true today. Flying drones are finding uses everywhere, from hobbyists to utility companies. Flying human transport drones—flying cars, that is—are right around the corner. And Openwater, Elon Musk's Neuralink, and a handful of other companies are developing brain-computer interfaces—not for the military to use to determine right from wrong, but to build more immersive connections between our technology and our minds. We could just as readily see this technology being used in an immersive video gaming system as in a military tribunal. Technologies of this kind offer promising civilian applications, whether to allow us to more easily learn, connect with friends and family, or relax in virtual worlds of video games or films.

This surely sounds bizarre at first blush. But consider that in 2017, 17.5 million people chose to have cosmetic surgery, including over 500,000 people who either had implants put in or liposuction to remove fat. An additional 860,000 people had even more invasive surgeries to replace hips or knees with artificial joints. Just a few decades ago, these procedures would have seemed bizarre and the statistics unbelievable. Considering the importance of human intelligence and the sheer amount of time we spend with our computers, it seems reasonable that large numbers of people would undergo elective procedures to enable cognitive superpowers or fully immersive VR experiences, once they are available.

And creating high-resolution lie detection would be less technically challenging than creating a high-speed bi-directional brain-computer interface. In other words, high-resolution lie detection should happen years before full brain-computer interfaces. So we should be thinking hard right now about the implications of high-resolution lie detection.

It's usually about this point in the discussion that someone says, "This is alarming. We should just make sure we block the military from using these technologies." I respectfully disagree. For one thing, even if the US Military decides it has no business playing around with brain-computer interfaces or transport drones, other militaries will, and so will many other industries. So we should at least think through very carefully how these technologies could be employed by less-democratic regimes, and what safeguards we should consider in defense as well as offense.

Furthermore, as these technologies become readily available, I believe their thoughtful application could represent an enormous leap forward for warfighting—and by extension, for humanity. Non-lethal and war-prevention technology has the potential to change what we think of as warfare, which should evolve from violence delivery into violence suppression.

I recognize that, for some, the scenario of the military swooping in, nabbing suspects, reading their minds, and locking them up sounds downright Orwellian, or just one shade too creepy. But I suspect that if you found someone from Raqqa or Mosul who had endured the US Military operations of the past decade, they would tell you that they would far prefer to see people getting carted off alive by military drones than getting killed by five-hundred-pound bombs dropped from the sky. "The real scandal is what's legal," the old expression goes. Apply that adage to the way we fight wars today: what's truly scandalous is what we accept as normal on our battlefields every day. What's truly disturbing are the things we've decided are a normal part of the evening news. Does that mean we let the military go forth unconstrained into a futuristic style of warfighting? Of course not. As with all progress in policing or the military, these things need to be done in a thoughtful and measured way and in as public a manner as possible.

In a way, the question of whether or not we ought to build more advanced weaponry is immaterial. The way our militaries are currently designed and resourced guarantees that we will develop more advanced fighting machines and weapons. It would have been impossible even a few decades ago, for example, to imagine that drones could encircle conflict zones, providing up-to-the-minute images and targeting ability. Those developments took place somewhat quietly and out of public view; it was military contractors and companies that produced the technology and the military that purchased it.

In a world that is already going to become more technologically sophisticated and roboticized, couldn't we point the direction of those efforts away from lethality? Couldn't we use automation and artificial intelligence to save lives on both sides of a conflict? If you take the long view and examine history, it turns out that's not such a radical idea. In fact, it's simply an acceleration of a trend that's already occurring. Consider figure 2, the chart of worldwide battle deaths.

This information comes from the Human Security Report Project, an independent academic research center. Anytime you're feeling especially down about the state of the world, it's worth reviewing their data. The chart shows that deaths from state-based conflicts have plummeted over the past eighty years.

While there were as many as 250 annual deaths per million caused by war in the 1940s, less than 10 annual deaths per million are caused by war today. High-intensity conflicts have gone down, and while some

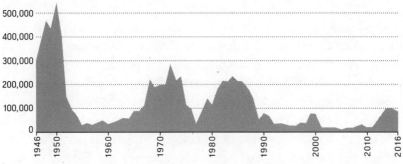

FIGURE 2: BATTLE-RELATED DEATHS IN STATE-BASED CONFLICTS SINCE 1946

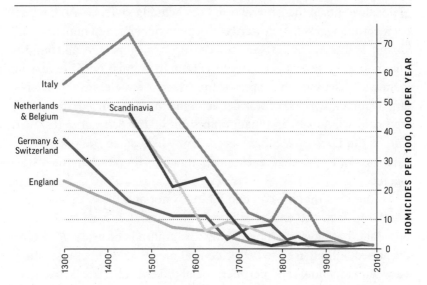

FIGURE 3: HOMICIDE RATES IN EUROPE SINCE 1300

commentators complain that the world is more dangerous today than it's ever been, that simply doesn't square with the data. And it's not only deaths from warfare that have gone down. Here's a fact you won't hear on the evening news: terrorism, genocide, and homicide numbers are down over the past hundred years, and in fact, on a centuries-long downtrend. Look, for instance, at figure 3, the chart of homicide rates in Europe over the past seven hundred years. See the steep decline?

Whether we are considering wars between nations or violent episodes between individuals, violence has dropped dramatically over the past few centuries. The decline in battle deaths around the world, then, appears to be part of a larger trend. For hundreds of years, large-scale violent conflicts have been one of humanity's prime killers. The death toll estimates for some wars (including World War II) are mind-staggeringly large. The idea that millions of people would be wiped off the face of the Earth because of geopolitical struggles is part of our very recent history. And yet, this aspect of our history has changed dramatically for the better.

In fact, looked at in a large enough perspective, even the bloody twentieth century—a century scarred by two world wars and multiple genocides—fits in the general trend of declining violence. Archeological records suggest that upwards of 15 percent of ancient humans died violent deaths. Yet in the twentieth century, war, genocide, and famine killed no more than 3 percent of the human population. One reason is that, as discussed in chapter 1, the advent of nuclear weapons has largely meant the end of war between the most advanced, industrialized nations. As Steven Pinker points out in *The Better Angels of Our Nature*, the Long Peace that has generally held since 1945 is only an extension of trends that reach much, much further back in the historical record.

So the idea that we can end killing is not at odds with historical trends. In fact, the end of killing is the natural extension of developments that have been underway for hundreds of years. As we've discussed, killing used to be the central means of conflict resolution between individuals and between social groups, from tribes to nation-states. But that practice has been on a welcome long-term decline, a decline that, I would argue, we can continue to accelerate toward zero.

Another reason that overall deaths from war have declined: better access to information about war's aftermath. You've probably seen the photo of a five-year-old boy from Syria, sitting with a bloody face and dusty limbs and hair on a foldable orange chair, next to medical equipment. The photo went viral in 2017. It brought home the reality of what was happening in Syria and how it was affecting real people. It personalized and humanized a conflict that, until that point, had seemed hopelessly distant and abstract to many.

That photo is one powerful example of a comparatively modern trend: access to information about lives lost, war's devastation, and the innocents harmed have made it harder to justify open warfare. It would be hard, these days, to hide a massacre or cover up a bombing raid gone wrong; there are too many cell phones within eyeshot, too many cameras, and too many social media networks to allow something like that to happen. That public pressure makes nation-states (and even non-state actors) less likely to inflict harm, kill, and begin conflicts. This is far from an absolute—there are groups that use social media to spread images of violence and to propagate hatred. However, the general impetus of more transparency is making violence more visible and less accepted.

World-ending weaponry and fear of public reprisal have combined to reduce the overall trend of global deaths as a result of war. While the story is not simple or tidy—devastating conflicts still take place to this day—the data paints a stark portrait of a world that is, in the main, becoming less violent and less likely to kill. One of the reasons I am optimistic about the future is that the past gives me cause for hope. The story of human civilization over the past several hundred years is one in which we've become more humane, less likely to kill, and more likely to treat killing as abhorrent. If this book makes any claim, it's that we ought to push that trend line further and faster—and that the newest era of technologies can help us do that.

I don't write those words to be flippant about where things stand with respect to weapons technology and lethality. Too many people on all sides of conflicts still lose lives. I am writing to add some necessary perspective and, dare I say, even hope. As radical as it might sound to consider a world in which geopolitical struggles are decided through less deadly means, it turns out we've already been building that radical new world for decades. Now all we need to do is accelerate and extend that trend line.

A CULTURAL INSURGENCY

ARLIER, WE ENVISIONED a scenario in which the Los Angeles Police Department is reconfigured as the LAVPD. It has a new focus and new technology that, in some cases and with the proper safeguards, allows it to proactively go after criminals before crimes are committed. What if we could imagine the same kind of shift in warfare and military technology?

Consider a scenario in which the United States, for whatever reason, needs to re-enter Iraq in 2045. Many of the same factions and fissures that made our modern war in Iraq difficult still exist; suspicion of US Military entry into Iraq is high. The conflict in Fallujah—a challenge that bedeviled US leaders from 2005 to 2010—continues to pose difficulty, with insurgents occupying the town, blending among civilians, and threatening American and Iraqi soldiers. The insurgency must be dealt with, but how? The last time the US Military entered this city, the fighting was long, bloody, and mired in complexity, with combatants intermingled with civilians. "Taking" the city meant owning rubble and a place riven with conflicts and violence.

Today, we might simply throw more resources at the problem. More troops, more barricades, more drone missiles. We might try to eliminate more enemy combatants, and pledge to hold on longer than they can hold out. But can we honestly expect that approach to work better than it's worked in the past?

Rather than trying the same thing over and over, and foolishly expecting different results, what if we radically reimagined our strategy? What if we tried to get better at violence prevention, rather than at lethality? What if American foreign policy looked more like the scenario I sketched at the beginning of this book? "Taking" an insurgent-held city like Fallujah or Raqqa might be possible without a single bullet fired, missile dropped, or life lost.

Just as with our LA police example, this scenario poses some self-evident challenges. How can we be sure that the non-lethal capacity of the drones is adequate and effective? What if they disable a child? What if they miss? What if their cameras go haywire and they start attacking US troops? What if they are hacked? All of these are valid and important questions, but I suspect that they are questions more about *how* an objective would be met than about *whether* that objective is worth meeting at all.

In my scenario, the objective is retaking the city with a minimum loss of life on both sides. That's congruent with how warfare has changed in general. During the Vietnam War, it became apparent that a fixation on "body counts" failed to adequately consider that there was no static number of enemy combatants waiting to be depleted by American arms, but rather a dynamic number capable of increasing as former civilians, perhaps seeking revenge, stepped in to fill the ranks of the dead. In modern conflicts, it is becoming clear that we cannot kill our way to victory.

Today, at least, we know better. Combat today isn't about racking up body counts. Few in the military would argue that point, and many military leaders make a gospel of the opposite: the military that can kill the least number is, in this century, the one that will be most effective at completing its mission.

In the introduction of a research report from RAND Corporation titled *Underkill*, the authors make the following claim:

> When the U.S. military is entrusted with responsibility for security
> in another country, the inhabitants should be accorded the same

protection from death and injury that Americans enjoy at home. A lower standard is indefensible on strategic, political, and logical grounds. In fostering effective and legitimate governments in such countries as Iraq and Afghanistan, the United States wants indigenous security forces to be as careful with the lives of their citizens as U.S. security services are with the lives of Americans. Because U.S. forces operating abroad must meet the same standard prescribed for indigenous forces, U.S. forces in such missions should be no more tolerant of death and injury among innocent civilians abroad than at home.

This is a controversial claim, but also an aspirational one that reflects a real shift in how the US Military intends to fight its wars.

There are several good and worthwhile reasons for this shift. International laws have made mass killing a crime, not simply a disappointing side-effect of conflict. Changes in international public sentiment in the wake of World War II and subsequent genocides make it harder for state actors to justify mass killings, and the moral cost attached to mass killings increased in the aftermath of the Holocaust and the Nuremberg trials. Changes in who constitutes "the enemy" also impacted how we think and talk about killing in war; today's conflicts are more often between states and non-state actors, rather than state-on-state. That means that the distinction between civilian and military is more nebulous, as was demonstrated in the many cases in the US wars in Iraq and Afghanistan in which the bodies being picked up after a firefight were dressed in civilian plainclothes, yet were firing weapons at US forces.

For the first time in human history, the most effective way to fight a war is whatever way will kill the fewest people yet achieve the objective. Marine Colonel Scott Buran, whom I introduced at the beginning of this book, is considered a national expert on these subjects. He was, like all Marines, trained in the art of killing. But over time, he also recognized that the goal of the US Military wasn't just to kill scores of people. "I go back to an old naval planning manual from 1936," he told me. "In it, they talk about the function and purpose of the armed forces. And they say, 'It's the reduction of the adversary's will.' It's Clausewitzian . . .

There's many ways to reduce an adversary's will. You can use lethal means but you can also use non-lethal means."

The military has tried to take seriously this idea that killing is one of only several ways—and often not the most effective one—to reduce an adversary's will. But if there's one institution in American life that is slow to change, it's the armed forces of the United States. In the Marines, the phrase "kill!" is used to mean everything from "hello" to "yeah, let's do this" to "okay." Our armed forces are trained to take life, and the tools they train with are primarily those used to take life.

But the fact that killing is a part of military culture doesn't itself make the institution of the armed forces unique; what does is the US Military's size and scale. The Department of Defense has a budget of over $700 billion. US Military spending is greater than the entire GDP of 90 percent of the countries in the world. Speaking of dramatic charts, have a look at figure 4.

Any way you slice it, the military's budget is big and growing. In practical terms, that means that the military has large contracts with big contractors that extend over many years. If you're a military contractor like Raytheon, the amount of time and money you invest into developing the latest-and-greatest weapon has to be made up for on the back

FIGURE 4: U.S. DEFENSE SPENDING

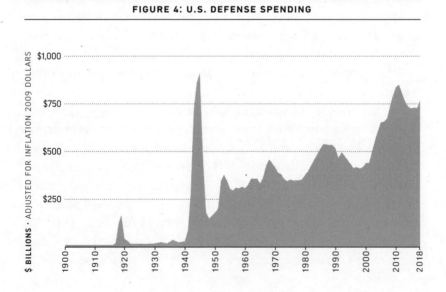

end. So you and the armed forces sign multi-billion-dollar contracts that last for some time, and you do not deviate from either what you promised or what you plan to build.

Changes to military hardware, let alone culture, take time. This is true both at the level of big military contracting and at the individual level. If you're a soldier on the front lines, you are prohibited from grabbing any weapon you can buy and bringing it to the battlefield. The US Military is a professional military force, not a ragtag militia. And that means that changing the weapons in soldiers' hands is going to be a long and cumbersome process.

Military contractors that make these weapons tend to be behemoths because they are dealing with the biggest and most complex bureaucracy out there: the Department of Defense. Navigating, negotiating with, and creating for such a big organization requires an entirely different skill set than that possessed by most entrepreneurs, which is, I suspect, part of why we haven't recently seen the kind of innovation in the defense space that we have in others.

But that doesn't mean we shouldn't try—not only because we've seen such dramatic decreases in military-related death in the last century, but also because a group of people within the military have devoted themselves to the possibility of non-lethal technology becoming standard issue for our soldiers, sailors, airmen, and Marines. Their name doesn't roll off the tongue: the Joint Non-Lethal Weapons Directorate, known as the JNLWD.

The JNLWD finds its origins in humanitarian conflict. I'm sure you're familiar with 1993's Battle of Mogadishu—if not by that name, then by the name Black Hawk Down, an incident in which two US helicopters were shot out of the sky by Somali insurgents during an American intervention in that country. In all, nineteen American service members died in the battle, and seventy-three more were injured—not only when the copters crashed, but also when their crews were surrounded by violent mobs, turning what was supposed to be a brief mission into an all-night standoff that resulted in American deaths and hundreds of Somali deaths.

Part of the reason for the standoff was that the US Military wasn't about to kill an entire mob that included civilians, even to save its own soldiers. There were too many civilians in the area, and it was much too heavily populated to use conventional lethal force without killing an unacceptably large number of them. The soldiers had no alternatives: the unfortunate choice was between killing hundreds or letting American soldiers get stuck in a terrifying and ultimately deadly situation. They were left with that choice, in large part, because there were few non-lethal alternatives to make a crowd disperse.

That problem stuck in the minds of the soldiers who were in Mogadishu at that time. Marine Corps Lieutenant General Anthony Zinni had introduced non-lethals during the first part of the intervention in Somalia, and less than two years after the battle, in 1995, he was responsible for getting the troops based there out of the country safely.

Zinni became a champion for non-lethal weapons, and the conflict in Mogadishu foreshadowed the increasing relevance of weapons of this kind on the battlefield. Zinni had the Army and Marine Corps work together to introduce non-lethal weaponry into Somalia—and thus, the Department of Defense's non-lethal weapons program was born.

The path that led General Zinni to focus on non-lethal weapons began before Somalia, in Vietnam. In 1971, he was wounded in Vietnam and was recovering in a hospital in Guam. Eager to return to service, he wasn't able to go back to Vietnam because of his wounds, but he did manage to get to Okinawa with the Third Force Service Regiment. He served as a guard officer, protecting the US Military base there, and he arrived at a particularly tense time in Okinawa: a combination of race riots and the actions of the Okinawan Communists kept him busy securing the base. "There were big crowds, demonstrations—it was pretty hostile," he told me.

Night after night, Zinni and his men would square off against protesters and demonstrators, manage big, unruly crowds, and deal with an opponent who might not have been firing bullets but was still putting lives and property at risk. "Some of it got pretty violent," he recalls. The soldiers had helmets and shields available as well as batons. Zinni trained with the local police to see what they used to subdue crowds; he also began to get creative. For example, he and his team filled high-pressure hoses with a dye that would take ages to come off the skin; if a crowd became too aggressive, the dye would mark anyone who was causing trouble—and later, his troops could pick those people out.

It was a crash course in non-lethal policing and operations, and Zinni carried it with him when he rose in the ranks and landed in Mogadishu in the aftermath of Black Hawk Down. "We ran into several sets of problems," he said. "For example, [Somalis] would demonstrate in front of our positions—the old burnt-out embassy in Mogadishu, for instance. When we would move our convoys out, there were people that massed and tried to take things off the truck, steal things. When we would take the trash and garbage out, people would swarm the trucks. We had intelligence that the bad guys were going to try to have little kids put explosives on the trucks, and we had a sergeant who shot a kid."

That particular incident remains in Zinni's mind. The Marines had received credible intelligence that their convoys would be targeted by children approaching with boxes containing bombs. In most cases, a child approaching a convoy was there to get something from the Marines: food, candy, and so on. But this also gave the guerillas in Somalia an opening: send a child up with a small package concealing a bomb.

It was a terrifying but effective way of getting close to the Marine convoys. One day, after a convoy went out, a child came running up to the truck with a box in his hands. And a Marine sergeant sitting in the back of the convoy was left with no choice but the worst choice possible: he had to shoot the child. The child was shot and killed. After the investigation into the shooting, the military concluded that the sergeant had taken the unfortunate but justified action. He had no other means at his disposal to stop the threat, which in this case happened to be a child carrying a box.

The Marine was crushed, his fellow Marines were shaken, and the incident highlighted for everyone the need for something other than violent weapons. "The Marines would come in and say we need some kind of capability," Zinni said. "Our only resources are our lethal weapons, and we don't want to use that ... The NCOs [non-commissioned officers] running these convoys were literally pleading with us. I had one Marine say to me, 'We don't want our only alternative to be an M16.'"

And it wasn't only convoys. Even basic security at the base would put the Marines in situations that tested their restraint. A journalist once told Zinni about watching a group of Marines guarding the gate as demonstrators protested outside. "She came to me after and said, 'I watched your Marine there. And there are these people screaming and yelling at him and it didn't even faze him. So I went up to him and asked him why he wasn't worried or scared. And he told me that he looked into their eyes and knew they weren't going to do anything.'" For Zinni, it was confirmation of what he knew from a lifetime of military service: the front-line soldiers did their very best to separate combatant from innocent, threat from bystander. "They're looking at women and kids," Zinni says. But their tools made no such distinctions; to an M16, everything was a threat to be killed.

THE MARINES were eager to find alternatives to lethal weapons, and in some cases, they tried to cook up their own. One example reveals how far the Marines would go to develop alternatives: General Zinni once left his headquarters and saw a few Marines and the headquarters company commander gathered around the open hood of a truck. He walked over and asked what they were doing. The Marines had figured that they might be able to attach a rod to the truck's battery, project the electrified rod from the front of the truck, and use it as a way of keeping the crowd at bay. "No, we're not going to do that. I could just see the CNN image now: Marines zap kid," Zinni said, and he put a stop to the project. But it showed him the great lengths his troops were willing to go to keep their lethal weapons holstered; this wasn't a situation with clear enemy combatants, and the last thing he or his troops wanted was civilian casualties in the name of keeping the peace.

Zinni canvassed all the military services for their best ideas on non-lethal weapons. What came back was disappointing. "The only service that had anything was the Air Force, and the only thing they had was pepper spray," Zinni says. The Air Force shipped over boxes of pepper spray, and as Zinni put it, "You had to wave it above your head three times, warn the person... it was ridiculous." Zinni and his team went searching for other answers. Some of the Marine reservists happened to double as California correctional officers in their day jobs back home. The prison system had struggled with these issues for a long time and it had developed technologies to help. The reservists showed Zinni and his troops rubber bullets, caltrops, sticky foam, and other experimental technologies.

The Marines used some of those technologies during the withdrawal from Somalia, and they helped keep tensions to a minimum. In fact, much of the tension came from outside Somalia—people began criticizing Zinni and his team for exploring non-lethals at all. A prominent journalist accused President Bill Clinton of forcing Zinni and the Marines to use non-lethals. "I responded to this, and I said, 'Bullshit. This is the Marines. They wanted this stuff. They needed it,'" he said. Another article from a top Army colonel—titled "Whither Sticky Foam?"—made a similar critique. ("For some reason, sticky foam became this big deal," Zinni remembered.)

Non-lethal weapons would, this Army writer argued, confuse the Marines. They'd be left not knowing when to be violent and when to stand down. "My Marines got all pissed off about this. We know when to use what weapon and when not to. Screw the Army," Zinni said.

The height of the public interest in non-lethals came when *60 Minutes* visited Zinni's base. The journalist was interested in the story and in the criticism that had been leveled at Zinni and others. Zinni and his Marines demonstrated the non-lethal weapons they had. The journalist was impressed but still skeptical. What about the criticism that Zinni was turning the Marines soft? The journalist felt he couldn't ask the question in that setting without biasing the answer; with Zinni, the commanding officer, standing around, his Marines would be unwilling to say that their boss was stripping the lethality out of their jobs. Zinni had a quick fix: "I said to the journalist, 'Go over under that tree. I'll take every NCO and officer away from the troops ... you ask them if they were confused about non-lethals or if they needed it.' And he did it." The journalist came back convinced that this hadn't been a subversive plot by Zinni, but rather a response to a genuine need among the front lines for something other than an M16.

ACCORDING TO ITS public mission statement, the JNLWD focuses on "non-lethal weapon capabilities [that] include blunt impact, marking, and warning munitions; acoustic hailing devices; optical distracters; electro-muscular incapacitation devices; and vehicle-stopping equipment." While the program was born in a peacekeeping and humanitarian context after Mogadishu in 1993, military leaders have since recognized the value of non-lethal weapons "in irregular warfare operations such as counterinsurgency, counterterrorism, stability operations, and counter-piracy."

Today, the JNLWD lives in a nondescript building in Quantico, Virginia, at the Marine Corps Base. In meeting with its leadership, it became clear that they are an insurgency of their own: a small group challenging the massive superstructure that is the US Military. Within that enormous enterprise, the mission of this small team is to ask: "Wait, what if we create weapons that don't kill?"

Importantly, the mission is *not* about replacing the lethal weapons that are in current use; it's to get credibility for the non-lethal options that are available today and to identify opportunities for their use. The staff of the JNLWD are optimists, but they are also realists. A former director of the JNLWD, Colonel George Fenton, told me, "One of the largest challenges I had ... was to get people to understand, it's not binary. It's not lethal or non-lethal. You've got to look at the situation and find the right capability."

A big part of the challenge is also making sure that units are trained to use non-lethal weapons. This can often be the primary obstacle to their deployment. "Military's got a fixed mindset. When you don't know what to do, you do what you know best. And you're only going to take into combat what you trained on," Fenton said.

To get any kind of support, Fenton had to outline cases in which non-lethals would be preferable to lethal weapons in theaters of combat. He'd break it down to as clear a picture as possible: "Let's take a guy on a patrol. A patrol moving through a town. You've got something over your shoulder that doesn't look right. Or you've got a sixth sense that's twitching. There's an action that takes place. I'd rather shoot first ... and let that first capability that's being employed be a ... device that's going to perform incapacitation. If he turns out to be a good guy, you've got one hell of a story to tell that night. If he's a bad guy, we got him. And then we can do some type of information exploitation on him. That's what I tried to get the military to embrace."

The JNLWD faces admittedly enormous challenges. Not only do they have to contend with the usual frustrations of military contracting and their small size relative to the US Military overall, but they also have to convince warfighters that the tools that have been in the hands of US soldiers, sailors, pilots, and Marines for years might need an upgrade. As Fenton put it, "I had one four-star general come up to me and say, 'What the hell are you doing with non-lethals? I'll tell you what non-lethal is: shoot 'em in the freaking kneecap. Get the hell out of that business. It's not what we get paid to do.'"

In another setting, briefing a high-level member of the Marine Corps, Fenton would be told, regarding his request for more money and a broader mission for the JNLWD, "Stop right there, Colonel Fenton. You, sir, are out of line. This program was never meant to be anything but a rubber bullet and beanbag program. You, sir, should be out playing golf." He was asked to leave the office of that high-level Marine, and as he put it, "I was about to go crazy."

Much of the work of the JNLWD is challenging a pervasive status quo—one that many people have both a financial and practical self-interest in preserving. Its leadership, in spite of those difficulties, has big ambitions. "We challenge all of our people to think of ourselves more as a start-up than as a traditional military structure," the former director of the JNLWD, Colonel Rey Mansinsin, told me. That kind of thinking has led them to do what might have seemed impossible even a couple decades ago: test and deploy non-lethal weapons.

ONE STORY from the JNLWD is simultaneously a stunning technological success and a tragic institutional failure. In the 2000s, the directorate created the Active Denial System. Think of the ADS as a heat ray, a weapon that could project a ninety-five-gigahertz blast of heat onto the recipient's skin and get them to move out of the way of the beam.

Conceived as a crowd control tool, the ADS did what it was supposed to do: if you stood in its way, you would feel a sizzling sensation on your skin until you moved out of range. It didn't kill you or maim you; it just increased your level of discomfort enough that you'd move. One of the test subjects described its effect like this to a journalist who covered the creation of the weapons: "For the first millisecond, it just felt like the skin was warming up. Then it got warmer and warmer and you felt like it was on fire... As soon as you're away from that beam your skin returns to normal and there is no pain." A smaller version was tested on a writer who described it as "a bit like touching a red-hot wire, but there is no heat, only the sensation of heat."

Sid Heal, a Marine Corps officer and one of the leading military experts on non-lethal weapons, provided this overview of the device: "The significance of this device is [first,] it was the most studied non-lethal device in history. We knew more about the physical effects of that device than anything else before it hit the market. The second thing: it was the first non-lethal device in history that provided adequate defense against lethal force, for the simple reason that the range exceeded lethal fire. We could target an individual beyond the range at which they could target us with lethal weapons without serious injury. These two things set it apart."

After being tested on more than ten thousand volunteers, the weapon was deemed safe, with a vanishingly low probability of permanent injury and a high degree of effectiveness. ("We had eleven thousand test subjects of which two were injured. And even those were slightly injured, second-degree burns, blisters," Heal said.) It was, in other words, the kind of non-lethal weapon that actually worked and that could be put to effective use on the battlefield.

The only real criticism leveled at the ADS from activists was that it worked too well. "The activists, militants, and extremists weren't complaining that it was dangerous. They were complaining that it worked so well that it would be overused," Heal said. "I answered honestly, 'Yes, that will happen.'... Why would you take a beating when you could subdue your adversary without it?" For Heal, and others who worked on the ADS, the real question was: What alternative did critics expect the military to pursue? The ACLU came out against the device, calling it a pain ray and advocating for roadblocks to its use. This was astonishing to Heal and others from the JNLWD. "They were comfortable with us using the same devices we've been using since 1820. But not Active Denial. Why would we spend forty million dollars and all this extra effort to save somebody's life? Needless to say, they don't have a better solution. They just have criticism," he said.

It had taken years and tens of millions of dollars to develop the Active Denial System. And it worked. It was even sent into the field.

During the late 2000s, the ADS was introduced into the battlefield in Afghanistan. And then just as soon as it was introduced, it was recalled. The reasons for this remain murky. Military leaders agree that it might have helped with what is known as "perimeter security"—the work of keeping military bases secure. Colonel Buran singled out a particular attack on American and British forces—a September 2012 raid on Camp Bastion—that might have been prevented with something like ADS. In that raid, two Marines were killed and many aircraft were destroyed by Taliban fighters who were able to breach the perimeter of a base. "[ADS] may have prevented that attack… not only saving national treasure and blood, but also those aviation assets," Buran argued. Even if that counterfactual can't be played out, it is a reminder that alternatives to lethal weapons for operations like perimeter security are uppermost in the minds of at least a few military leaders.

ADS's deployed-but-never-used outcome should be a source of concern to us all. David J. Trachtenberg, a former principal deputy assistant secretary of defense, called it a missed opportunity and commented that "the non-lethality of the ADS system could prove useful in a counterinsurgency operation where avoidance of civilian casualties is essential to mission success." And yet for all that—a weapon that worked, with use cases and testimonials from members of the military—nothing came of it. Colonel Buran captured the sentiment of many in one statement about Active Denial: its lack of use, he said, "is shameful."

When I asked him why the US Military leadership recalled the ADS and it was never used in the field, Sid Heal had a one-word answer, "Cowardice. It's as simple as that." Heal continued, "A secretary-level military leader decided against deploying ADS on the battlefield, saying, 'Non-lethal weapons such as high-power microwave devices should be used on American citizens in crowd control situations before they are used on the battlefield. Domestic use would make it easier to avoid questions in the international community over any possible safety concerns.' Meaning, if we're not willing to use it here against our fellow citizens, then we should not be willing to use it in a wartime situation, because if I hit somebody with a non-lethal weapon and they claim that it injured them in a way that was not intended, I think that I would be vilified in the world press."

Heal paused before concluding, "So the option that's left with us is that we drop bombs on people. Military leadership hid in the sanctuary of anonymity, and waited for someone else to be first, because nobody wants to take the risk to be first. There is no doubt in my mind that, had they been able to use the Active Denial System, lives would have been saved." It's a damning assessment from someone who spent decades in a position to explore and weigh the alternatives. "The standard is not perfection," said Heal. "The standard is the alternative. If you take away whatever we're using that's effective, you need to be accepting of the fact that we're still going to have to solve the problem with more primitive options."

It's astonishing that even people who are well versed in military affairs and foreign policy barely know about this technology and its effectiveness. Do we think that the Black Hawk Down incident—with all of the suffering and loss of life it entailed—can't repeat itself? Imagine the scene, say, two years from now: American Marines are stationed in Damascus, helping to preserve a fragile peace after the end of the Syrian civil war. Imagine that, once again, an insurgent cell gets its hands on a rocket launcher and, once again, a Marine helicopter is unable to evade the blast. Imagine the scene from Mogadishu playing out once again, as if in a recurring nightmare: the injured Marines holding off an enraged crowd with small arms, the hours-long standoff as backup troops maneuver to extract them, the dead and wounded on both sides, the sickening images as dead Americans are dragged through the streets.

The first time, in Mogadishu, it was a tragedy. This time, unless the American military changes course, follows the lead of General Zinni and the JNLWD, and takes non-lethal technology seriously, it will be something far worse than a tragedy, because the very tools that are capable of preventing it are available but sitting on the shelf.

THE CHALLENGE here isn't solving the tech problem; in some important ways, it's already been solved. The challenge is in updating our training, our standards, and our general approach to war to embrace the newest technologies. That's not a question for R&D labs; it's a question for warfighters and strategists. General Zinni pointed out that questions about weapons tend to come down to budgets. "The way military budgets work, it's a competition. [When] you're competing with lethal weapons systems, [non-lethals don't] get the kind of constituency and support that the others have. It's kind of like, 'Yeah, that's fine. Great. If we have to decide where to put our money, it's not there.'"

But Zinni did offer one way forward. "The case has to be made that use of non-lethals is not something separate, but is adding to your kit and how you integrate it into mission accomplishment along with lethal systems." What Zinni and others argue is not that the gun should be taken away, but that it should be augmented with non-lethal alternatives that allow someone on the front lines a choice about the appropriate force necessary to deal with a situation.

Watching the patterns that have emerged in the military conflicts of the twentieth century, it has become clear that the lack of non-lethal options is more than an inconvenience. The lack of non-lethal capabilities has become the weak link of the world's most powerful military. Adversaries of the United States force American troops into unwinnable situations—such as a child approaching a guard post holding a box that may contain explosives. In such a situation, there is no good choice. Not firing puts US lives at risk. But killing a child is sure to inflame anti-American sentiments in the local population, drawing ever more radicalized people to join the fight. For just the same reason, insurgent forces often take advantage of "human shields" or hole up in schools or hospitals.

Another difficulty is that international laws of war are quite strict about using non-lethal weapons on non-combatants. Those non-combatants are often drawn into situations that aren't clear fights but that can be dangerous. The inability of our soldiers to use non-lethal means puts those lives at risk. "The 1957 conventions predate everything we have available to us. Maybe they should update the policies?" Heal said. "Not being able to [use non-lethals] meant that we had to subject civilians to conventional weapons. One of the things they'd do in Somalia is that they'd drive the innocents in front of them, because the Americans had such a great reluctance—even a revulsion—to target these victims. But it offset our technical advantage because it allowed them to get in range and attack us with conventional weapons."

The greatest militant weapon against modern militaries is to draw their lethal firepower into use against innocents, and to ensure that the results are broadcast as widely as possible—both to enrage the local population to build resistance and recruit more militant fighters, and to undermine the moral authority of the operation and erode support back home.

If you'll recall, one of the things that's inspired me throughout my career has been sci-fi applications of non-lethal weapons technology. The *Star Trek* phaser is an obvious inspiration for the TASER weapon; the stun feature on the guns in *Star Wars* is another. In both cases (and others in popular culture), what is at stake is the future of the galaxy. Groups are at war, and weapons that incapacitate but don't kill play a central role in trying to win that war. There's no reason why our reality, in this respect at least, can't look a bit more like science fiction.

I believe the curves we discussed earlier in this chapter can continue their downward march and that we can do even more to accelerate that trend. I believe the strategic imperatives of combat are going to lead us away from killing and toward missions of disabling, capturing, incapacitating, or neutralizing threats without taking life. Not only will this be in our strategic interest, but it will also make us more effective actors on the global stage.

But that progress needs to be accelerated, and at the moment, I worry that we may be going in the opposite direction. In 2014, the US Army War College shifted its curriculum, dropping all courses on stability operations in favor of those focusing on "traditional warfare." In 2018, the US Army Peacekeeping and Stability Operations Institute (PKSOI) was notified by the secretary of the Army that it would be closed or severely diminished in capability. In 2018, the National Defense University closed its Center for Complex Operations, and Penn State University, which has taught Non-Lethal Weapons in various military programs, was informed that its contract would not be renewed. Some sources even shared that the US Marine Corps was actively considering whether to divest itself of the Joint Non-Lethal Weapons Directorate.

I find these trends and attitudes concerning and regressive. The current international wars have been long and frustrating, without a clear path to victory. This has triggered a strange nostalgia for the wars of the past, which, although far bloodier, ended with decisive victories. What benefit do we gain by building ever more lethal tools for traditional warfare, or by teaching today's soldiers-in-training that they ought to go into situations of moral and ethical ambiguity with only one goal: to kill the enemy?

Nations tend not to have the luxury of choosing their next adversary in war. If recent history is any guide, the adversaries who attack the United States are unlikely to be nation-states against whom a traditional land invasion will be effective. It seems that we are shrinking from the current challenges, hoping to return to a more predictable past—as though building war tools for yesterday's wars will somehow shape future conflicts to fit our military desires.

I would appeal to our national military and political leaders to look to the future with sober eyes and focus on the resources that will enable the protection of the national interest from the most probable future threats. Imagine a non-lethals program that took on the size and scope of the Manhattan Project. With the right focus, energy, and resources, our scientific and industrial leaders can accomplish amazing things. Developing the tools to stop pregnant women and children without killing them is surely an easier problem to solve with today's technology than it was to create nuclear weapons that could incinerate entire cities with the technology of 1945. These problems are solvable. We simply have to apply the willpower and know-how to solve them.

I often think of the well-worn images we have all seen from footage of the end of World War II. These grainy videos show US troops greeted as liberators. Soldiers traveled through far-flung, war-ravaged towns in France or Belgium, and they'd be met by thousands of people, flooding town squares and streets, throwing flowers at the soldiers and hugging them, some with tears in their eyes. Anyone who watches one of these scenes now can't help but be moved: they are images of the US Military at its best. If we have a deep wellspring of respect and reverence for our military, that's because our soldiers are, at moments of great international peril, instruments of incredible good in the world.

These scenes are fewer and further between now. No one is hosting parades for US soldiers in Kabul or Baghdad. There are countless reasons for this, of course—some that we can control, and others that we can't. But my prediction for non-lethal weapons is that they can help us win back the "hearts and minds" of those who, today, might be suspicious of our motives and methods. As one former director of the JNLWD put it, when talking about the changes necessary to defeat an enemy in the modern era: "Understand their culture. Understand their mindset. What are their customs? What are the things we need to be able to do, understand, and respect so that we can connect with them? You don't get that with a drone strike."

Retired Marine Corps Lieutenant General Buck Bedard put it more bluntly to me, "Often times what's difficult today, especially when you're not dealing with a nation-state but you're dealing with a terrorist threat, is sorting out who the actual enemy is. And sometimes to do that you've got to immobilize people or you've got to knock them down and then sort out who they are. And I think today, too many times, we sort them out by killing a bunch."

Could new technology help us be greeted not as conquerors but as peacemakers? I can't say for sure—but I know there's ample reason to try. Let me end with the best case we can possibly make for trying to advance the technology of warfighting: the words of a warfighter himself, a story from Marines in Iraq.

> I was in Iraq, interviewing an Amtrak [armored vehicle] driver. He was almost in tears. I say it was hard to talk to him because he kept choking up. I'd never seen a Marine in combat choke up so profoundly. I interviewed five guys on his team, and three of them were having a difficult time telling the story.
>
> The day before I interviewed them, they had had a vehicle approach their checkpoints. It was very fast, coming down the road. Needless to say we were already experiencing suicide bombers and that was the first thing that came to their mind because that is the threat that they're going to have to deal with.
>
> The platoon commander, a lieutenant, ordered warning shots. So they fired their machine guns. I don't know if you've ever seen

warning shots but if you're in a car they're very ineffective. The only thing you can usually hear from inside the car is a distant popping sound, I mean it's not startling at all. Even though ... they're shooting and having ricochets off the ground [and] everything. The only thing you really see is faint puffs. It's not like on TV where you have these dramatic effects, it's very ineffectual as far as a signal. So the vehicle just kept coming. So they ordered it again. I believe they fired two series of warning shots.

The third time, though, they [were] told "light it up," and they fired into this and basically destroyed this SUV. Then they waited for two and a half hours for EOD [explosive ordnance disposal] to come and clear the device so that nobody got killed by a car bomb, which is what they thought was happening. But what they got was a family, and the family ... looks like they were just escaping from the Iraqi side and trying to get out. And there was a little girl alive in the backseat but she had been gravely wounded.

Everybody else was already dead. The men telling me the story could barely talk to describe their feelings when they saw it was a family and not a suicide bomber. And then they saw the little girl. Needless to say, they did everything they could. They called helicopters and medevac and corpsman. But she died.

This guy is going to live with that the rest of his life. He was the machine gunner. He could barely tell me the story.

JUDICIOUS
SURVEILLANCE

THE MORNING OF March 22, 2016, started out as a day like any other at Brussels Zaventem airport. Passengers waiting in the security screening line displayed their IDs and boarding passes, took off their shoes and belts, and pushed bins of their belongings into the X-ray scanning device.

At 7:55 a.m., an explosion ripped through check-in row eleven, hurling nails as shrapnel through the airport. Nine seconds later, as panic set in and people realized what had happened, a second explosion hit just a few rows away. An hour later, as the country came to grips with the tragic events of the morning, a third bomb was detonated in the Maalbeek Metro station in downtown Brussels near the European Commission headquarters. Less than two hours after it started, the series of attacks had left thirty-five people dead and over three hundred injured—and an entire nation and continent scarred.

Images from closed-circuit television (CCTV) footage at the airport showed the terrorists casually pushing carts with the explosive-laden suitcases through the arrivals lounge. Far from being anonymous or unknown individuals, the five men involved in the attacks were well known to police, belonging to a terrorist cell that had been involved in the November 2015 attacks at the Bataclan concert hall in Paris.

One of those men, Osama Krayem, had been fighting for ISIS for years; even before the bombing, he was one of Europe's most wanted fugitives. The police had his information, including photos. Another, Khalid El Bakraoui, had three warrants outstanding for his arrest. The others had either served prison time or they were suspected of having traveled to Syria to fight for ISIS. In other words, these are people for whom the police had paper trails and information available.

But the fact that authorities knew of these men and were looking for them made no difference on the morning the men acted out their plans. Looking at the CCTV footage from the airport, and knowing of the extensive CCTV coverage across the public transit systems throughout Europe, I couldn't help but wonder when the technology for biometric identification—whether from facial recognition, gait recognition, or even behavioral threat detection—will reach a point of maturity where it could make that system-wide knowledge available in the moments when it could save lives.

Police across Europe were looking for these men, and yet they sauntered right across the view of myriad security cameras, while they were pushing bombs on carts in an airport. And no one was able to piece the information together until well after the damage was done.

It's not enough to think about new technologies in one particular domain of self-protection or police work. Part of the thought exercise of this book is thinking through the broad applications of new technologies in all domains that deal in life, death, and security. So it's worth turning our attention to the large and sprawling homeland and border security system, a system that, as this story illustrates, could surely benefit from the high-tech advances of our era.

Here in the United States, the prevailing narratives around facial and other biometric recognition systems are focused on the risks of abuse by the government. We fret endlessly about all the ways that agencies in Washington might misuse such technology. Again, I don't mean to dismiss those concerns. But if something like the attacks in Brussels takes place at JFK or LAX, I suspect we will all have a significant and sudden shift in our outlook, and the changes we institute will be abrupt and potentially ill-considered.

Many hurried changes took place in Washington in the immediate aftermath of the September 11 attacks. A whole new bureaucracy spun up virtually overnight to deal with the threat of terrorism; only in recent years have we learned just how vast that apparatus was and how much information it was sifting through, in some cases without the public's knowledge or assent. That moment should teach us a valuable lesson: it is better for us to evaluate the cost versus the benefit of these kinds of surveillance and analysis technologies before tragedies strike, not in their overheated, emotional aftermaths.

Some of the slapdash, rapid expansion of our system of border and homeland security could be excused. The country was still reeling from the shock of the attacks, and at the time, no one knew if another tragedy was just around the corner. But in the many years since, have we ever stopped to seriously consider if the efficacy of this effort matches the size of the bureaucracy that was created around it?

THE PUBLIC acceptability of various security technologies will, of course, vary by country. There is a disproportionate concern about facial recognition technologies in the United States. In Europe, government officials have admitted to me that there's more public acceptance of biometric identification technologies after years of terrorist attacks in several major metropolitan areas.

Some of this difference is, no doubt, cultural. I've seen this first-hand with the deployment of body camera technology. In the United States, the primary driver of body cameras was to give assurance to the public that agencies would have more oversight over officers' behavior. Put more directly: the impetus for the cameras was to protect the public from aggressive police. In Europe, this has not been a major factor driving the adoption of body cameras. There, the argument in favor of cameras is to prevent another knife attack or to respond quickly if a truck mows down a crowd.

While the public concern about surveillance technologies poses one class of challenges, there is another that is, perhaps, more difficult to overcome: government inertia. There is an argument for taking our time, being deliberate, understanding costs and benefits, and designing safeguards. These technologies deal with sensitive enough information that we ought to work hard to get it right. But in many cases, the delays to adoption have nothing to do with getting it right. They often have to do with the inevitable intransigence of government bureaucracy.

Take, for instance, the problem of scanning luggage at airport check-points. Today, we've all agreed to subject ourselves and our luggage to fairly invasive screening at security checkpoints before we get on an airplane. We have made the trade-off that we will allow representatives from the Transportation Security Administration (TSA) to peer through our luggage and even through our clothes, in order to ensure that no one is taking a bomb or a weapon aboard an airplane. That seems rea-sonable enough: following the events of 9/11 and subsequent failed attacks, I am more than willing to submit myself and my belongings to a search designed to keep me safe.

The problem isn't the invasiveness of the search, though, it's the ineffectiveness of it. While the airport screening process is pretty adept at finding water bottles and nail clippers, it turns out it is not very good at finding less common but vastly more dangerous items, like guns and knives.

In 2015, agents from the Department of Homeland Security posed as passengers and put bags through the TSA screening system. Those bags contained fake explosives and banned weapons. Out of seventy tests, the TSA agents failed sixty-seven of them—over a 95 percent failure rate. In an astonishing moment, a Homeland Security official posing as a passenger actually set off a magnetometer scan. The fake passenger was stopped. But the officials who patted him down missed the fake explosive taped to his back. These failures led to the removal of the head of the TSA.

This is a stunningly poor performance, but perhaps more important for our purposes, it's also an inefficient one in which both the tech-nology and the process are to blame. And let's not forget the general unpleasantness of having to subject yourself to a TSA screening before you board a flight. Not only are we spending time and money for a secu-rity system that appears to fail most of the time, but we're also left with an endless line of unhappy travelers enduring extra screening proce-dures for limited benefit.

One person became obsessed with this problem and even developed a solution. When you learn about it, it feels like an obvious improve-ment that we ought to deploy immediately—and yet, it never made it past the government bureaucracy. Ian Cinnamon is a wunderkind—a twenty-seven-year-old MIT graduate who studied brain and cognitive science and started a company called Synapse Technology Corporation to develop AI-enabled technologies to help make security checkpoint screenings more effective.

Cinnamon identified that the failure of checkpoint screenings isn't because of lack of effort on the part of security officials. It's that the human brain isn't built for repetitive visual search tasks. That's especially true when what you're looking for—guns, knives, bombs—appears on the screen very rarely and other images—suitcases, backpacks, laptops—appear regularly. It's what's called "sporadic visual search," and as Cinnamon explained to me, "If you or me or anyone is staring into the world around us, and you're looking for a specific item or target, the research shows that when the target appears infrequently, your brain doesn't light up. The opposite happens: you don't even notice it. You might be standing on a street corner looking for a specific make and model of car. The thing drives by right in front of you. You might see it. But you're not able to recognize that you've seen it. It's just not being processed by the neurons, axons, and synapses in your brain."

Humans have a very difficult time reacting quickly and accurately to something that happens so infrequently, and they become attuned to the things they see all the time. Fatigue and bias also cloud the visual inspection. Cinnamon believed machines could do better—they aren't as susceptible to boredom or bias, and they don't get tired. So he built one and proved it.

The same scanning technology could be used to help doctors better identify cancers and TSA agents better identify weapons. Cinnamon decided to focus on the security research angle. He found the cancer angle interesting, but in medicine, unlike at security checkpoints, people could go to another doctor for a second opinion. Or they could take a second round of screenings to see if, in fact, they had cancer. At security checkpoints, there was no margin for error.

"This is a problem of a human mind and brain, and no matter how well-trained you are, you're going to have the same problems and limitations caused by the sporadic visual search," Cinnamon said. "What we realized was that artificial intelligence is the perfect tool to solve this problem. It's a constrained problem space. AI is modeled after the human brain. So you can train an AI algorithm on ten different makes and models of firearms and then show it a firearm it's never seen before, and similar to a human being, it's able to generalize and understand that this firearm it's never seen before is actually a firearm... Unlike the human brain, though, the algorithm doesn't get tired and doesn't fatigue."

Cinnamon's company got to work. Using AI algorithms, Synapse demonstrated 98.73 percent effectiveness in identifying firearms being screened, with a false positive rate of only 0.01 percent. It was an astonishing, potentially transformative result. But when Cinnamon showcased his results to one security screening agency, he was surprised at the response. He was told, "So, you're asking me to buy a product that we know will let 1.27 percent of guns through? No way." To his shock, the discussion was over: his AI-based system was a nonstarter, even though it outperformed human screeners by a mile.

He was stymied, in part, by the age-old problem of any bureaucracy, particularly big bureaucracies in government: the best and newest idea can fall victim to a long, slow contracting and testing process. The two companies that dominate this space had long-standing, extensive, complicated contracts with the government—and they had little interest in listening to a small start-up that was telling them how to do things better. Those companies had also mastered the procurement, RFP, and proposal process that the government had set up, and they had passed multi-year testing for their products. All of that conspired to make it difficult for someone who invented something new, even if that something new outperformed what was currently in use.

Cinnamon's story also illustrates a truth about how technology and human beings interact: in many cases, it is difficult to get human beings to trust technology over their own agency and decision-making. So when it comes to scanning luggage in a TSA line, the human psyche believes it can outperform a machine—even if we are statistically far, far worse at it. The old story of John Henry, the "steel-driving man," defeating the steel-driving machine is a powerful one: human beings want to believe we're better than the machines.

Synapse is working on these ideas still, and Cinnamon continues to push forward, working to educate the market on the vast improvements artificial intelligence can provide for security.

LET'S RETURN, for a moment, to the attack in Brussels. Let's say, for the sake of argument, that AI-powered video technologies *were* able to identify the threat. What happens next? If you have ever traveled through a major European airport, you have seen what the most likely response mechanism would be: a swarm of soldiers appear armed with assault rifles. We all may feel more comfortable knowing that there are armed soldiers at the ready in a situation like this; however, while these armed human assets create a powerful deterrent effect, are they the right response in a crisis situation of this kind?

One of the reasons to treat border and homeland security as distinct from something like a military operation is because the most difficult homeland security operations tend to occur very rapidly and with civilians around. In Brussels, even if those would-be attackers had been identified, it's not clear that firing a high-powered assault weapon in the arrivals hall of a crowded airport would have been effective in neutralizing the threat. And that's ultimately what a response ought to do in this situation: neutralize the threat.

I would argue that a non-lethal solution exists even in these ticking-time-bomb type situations. That's in part for the same reason that we'd like a non-lethal array of options in combat overseas: because in an ideal world, we'd have technology available that helped us capture potential suspects. Abroad as well as at home, a dead would-be terrorist can't tell you anything about their organization, their sources of funding and training, and whether additional attacks are planned or already in progress. They can't point you up the ladder to more important figures in their organization. They can't reveal that there's a third bomb in Maalbeek Metro station.

Beyond the loss of information that comes from killing a suspect in a situation of that kind, we know that killing creates martyrs—figures that radical organizations hold up as the face of their movements and use to inspire new cycles of violence. That's why suicide bombing is doubly effective as a terrorist tactic: it creates both victims and heroes. And that's why it's best to avoid playing into terrorist hands, by whenever feasible finding alternatives to killing would-be terrorists.

Add to all of that a fact that I've already emphasized in this book— killing is traumatic even for the most professional law enforcement officers. Study after study shows the trauma that even justified killing can cause to professionals. I'd urge us to consider how new technologies can affect not just identification of threats, but also the resolution of them. Our borders—and airports, train stations, subway stations—would become more dangerous, not less, if we could identify every threat but the only tool for responding to those threats was a hail of gunfire.

ONCE A suspect is subdued, another set of problems presents itself: how to learn as much from them as possible in the shortest time possible. A decade or two ago, I think it's fair to say, those problems weren't quite so acute. Terrorist groups like Al-Qaeda were more or less structured as top-down institutions. They had a clear chain of command, a well-defined leadership structure, and a set of objectives. The national security agencies could often pick up word of their plans by intercepting electronic communications or tracking human intelligence. That intelligence might not be detailed enough to pinpoint the time or place of an attack—but if an attack took place, agents would have an easier time connecting the dots between the suspect and the larger organization.

Well-organized terrorist groups can obviously bring a great deal of funding, resources, and know-how to bear on the cause of sowing death and destruction; a "lone wolf" couldn't have planned and executed the 9/11 attacks. But the trade-off for these top-down groups is that what they gain in organizational power, they lose in their visibility to their opponents' national security apparatuses. In other words, you can't be big and organized without leaving a footprint. It was that footprint that allowed the dismantling of Al-Qaeda after 9/11, up to the killing of Osama bin Laden. Years of footprints had left a trail, and US intelligence and military followed it to the head of the organization and many others beneath him.

Things are different today. We're seeing a rise in "homegrown" or lone-wolf threats—killers and would-be killers who don't belong to any organization. They might not report to any leader at all, they might not communicate their plans to anyone, and they are unpredictable in a way that larger terrorist organizations are not. They may have been radicalized through social media or YouTube—one day, they're one of millions of anonymous users of the internet; the next, they're a lethal threat. Lone-wolf terrorists hide in plain sight. That makes it hard to anticipate their plans and to foil them before they take action. And in the event that their plans are foiled, they're more difficult to plug into an existing schema of intelligence.

Those interrogating a lone-wolf suspect aren't so much looking for the suspect's place in a larger organization as they are looking for patterns of behavior. They want to understand the ideology that motivates them, the sort of media or information they consumed on the way to radicalization, and the kind of help they had, if any. Of course, they also want to know if any friends or compatriots are planning similar attacks. But we know that such suspects are often unwilling to cooperate; in many cases, they were already willing to lose their lives, and they have little to gain from cooperation.

How do we identify those patterns? How do we make our way through a thicket of information to determine what, if any, threats to the homeland can be gleaned from a massive amount of data? Sadly, we know that many governments, including our own, have turned to torture in such cases. But not only is torture a gross violation of human rights—study after study has also shown it fails to produce accurate information. And that makes sense. Someone suffering intense pain is likely to say whatever it takes, and whatever the interrogator wants to hear, to get the pain to stop.

That's where technologies of the future can come in and help us prevent terrorist attacks without having to declare war, fire a bullet, or invade a foreign land. There are a number of promising options for conducting accurate interrogations that don't cause pain and don't violate suspects' rights in the same way.

None of us wants the government rifling through our Facebook posts, our photo libraries, or our text messages. We have justified concerns about the government tracking our political involvement, our social lives and sexuality, or our financial histories.

At the same time, we are outraged and dumbfounded when the government fails to see the flashing red lights that often precede massacres. On February 5, 2016, Nikolas Cruz was pictured with guns on Instagram, and a neighbor's son reported to the sheriff's office that Cruz "planned to shoot up the school." Over a year later, in September 2017, a blogger in Mississippi warned the FBI that someone named Nikolas Cruz wrote on his YouTube page "I'm going to be a professional school shooter." Several months went by before Cruz took seventeen lives at a high school shooting in Parkland, Florida, on February 14, 2018.

If better social media monitoring could have prevented this tragedy, would that be worth it? Or think back to the chapter 5 discussion of violence prevention—if better social media monitoring could have prevented the Pittsburgh synagogue massacre, would that be worth it? More broadly, how can public safety organizations use data to protect us without infringing on our privacy in inappropriate ways? I know how hard the question is, and I know that the most thoughtful answers aren't going to fit neatly on a bumper sticker. But surely the greatest minds in technology and policy can come up with solutions that make us safer than we are today without sacrificing all sense of privacy.

THE BROADER point is that we have, until this point in human history, had imperfect tools to keep nations safe and borders secure. All too often, the absence of reliable tools to subdue threats and discover their origins has led to a reliance on unpalatable options, like torture.

In the near future, however, advancing technology might allow us to secure our borders and protect the homeland without having to take life or commit human rights violations. (At least in the traditional sense: these new technologies may enable an entirely new class of privacy and search discussions.) These technologies aren't perfect, and they carry their own risks and cautionary labels. There are real and genuine concerns about privacy and the use of personal information, and the heated debates about those matters are heated for a reason: how, when, and why we gather data and analyze it in order to keep the country secure are subjects of sensible concern to policymakers and the public.

As I've argued throughout, if we evaluate new technologies against a standard of perfection, they are all bound to fall short. But if we evaluate them against the practices that governments have used in the past and may well use again in the future, they come out looking far better. We can't hope for a world without ethical quandaries, any more than we can hope for a world in which no angry young person ever gets radicalized and decides to kill. What we can hope for is a world of incremental but real improvement—a world in which the work of keeping us safe is far more likely to stop, rather than perpetuate, cycles of violence and killing.

JUSTICE MODERNIZED

I**F YOU WERE** like many millions of Americans in 2014 and 2015, the criminal justice system entered your life in an unexpected way: as entertainment. Those were the years in which dramas like Netflix's *Making a Murderer* and NPR's *Serial* made their debut and became smash hits. They left audiences both riveted by the drama of true crime and dumbstruck about how the criminal justice system determines guilt or innocence. For many people, I suspect, it was the first time they wondered: What happens when judges and juries get things wrong?

Making a Murderer took a full decade to complete. The series chronicled the story of Steven Avery, who served eighteen years in prison for sexual assault and rape. He was exonerated by DNA evidence in 2003, only to be re-arrested and convicted later on a different murder charge. Avery had gone after the police department that had wrongfully convicted him the first time, and the documentarians illustrate how his pursuit of exoneration might have led the police to pin a new murder on him.

They follow the ups and downs of Avery's case, with special attention to the coerced confession of Avery's nephew, who police alleged worked with his uncle to carry out the murder. There wasn't much evidence for that, and the boy had obvious mental difficulties, the kinds of issues that would have made it difficult for him to tell police interrogators exactly what his uncle might have done and when.

In the documentary, we witness the cops grill the nephew for hours, and before long, he confesses to the crime and reveals his uncle's alleged participation in it. For many people, these were troubling scenes: viewers' blood boiled at the footage of a young man pressured to confess to a crime that he seemed to have no part in, at least until the police created a storyline for him that included his participation. The intensity of that feeling—as well as the public response of critics, celebrities, and legal experts—drove *Making a Murderer* to popular success. And more than that, the public outcry about the series even led a judge to hear an appeal of the case.

NPR's *Serial*, which debuted in late 2014, holds the world record for most podcast downloads (as of this writing, over 175 million). The podcast explored the murder of Hae Min Lee, an eighteen-year-old student at Woodlawn High School near Baltimore, Maryland. Her boyfriend, Adnan Syed, was convicted of killing her and sentenced to life in prison. But the journalists behind *Serial* brought to light that Syed's defense lawyers had, apparently, failed to defend him. There was evidence in the form of cell phone records as well as an alibi from a friend of Syed's who was with him at the time of the murder—and who was nowhere near the site of the killing—that should have exonerated him. The defense attorneys failed to use that evidence properly; *Serial* brought those facts to center stage. (Syed's conviction was overturned, and he's being retried.)

Making a Murderer and *Serial* reached audiences that may have never found themselves on the wrong side of the American justice system, and it shook their faith in that system. How is it, many people asked themselves, that our justice system could seem so arbitrary and capricious, in the United States, in 2014? Surely courts, judges, and juries are supposed to prevent wrongful convictions? And if they don't, how do we fix what seems like it ought to be a fixable problem?

If you have candid conversations with people on the inside of the system, as I have through my work with police departments around the world, you'll find that they sometimes share this lack of faith in how justice is administered. Finding out that innocent people get put away—or that criminals they've risked their lives to apprehend get off on technicalities—can be utterly demoralizing to the men and women on the front lines of the justice system.

Until this point, I've argued how technologies will reduce and even eliminate sanctioned killing. But I also want to explore the what, how, and why of technology's impact on the moments just after an arrest is made. Because just as non-lethal technology and advanced computing power and information technologies can decrease the likelihood of a violent event, those same advances can change how a crime is dealt with after it takes place.

I also believe that a more effective criminal justice system will act as a deterrent on crime. If we could guarantee that 100 percent of criminals would be caught and punished, then crime rates, I suspect, would drop dramatically. But our current justice system is deeply ineffective. Cases are often dropped because they're "not worth it" (too much cost), criminals get off because the district attorney couldn't build the case against them (not enough evidence), and in the most horrifying cases, the wrong person is found guilty and punished.

If we could repair these issues, not only would justice be served, but crime itself would also be reduced. We'd be tackling both ends of the problem. Let me put it differently: to some extent, today's criminals place a bet that they can get away with their crime. The harder we make it for criminals to get away with crime, the fewer would take that bet.

I believe technology can make it easier for us to create a more effective justice system. That isn't a new or revolutionary idea. Our criminal justice system has always adapted itself to new kinds of evidence-collecting methods and information-gathering practices. As early as the 1900s, courts and legal experts were struggling to determine how much weight and credibility to assign to the most primitive lie detectors. Fast forward to the 1980s and '90s when states began to require criminal offenders to submit DNA samples.

The use of DNA in the criminal justice system is a useful case study in thinking about the potential of—and potential objections to—the introduction of new technology into how we adjudicate crimes. When the examination of DNA was first introduced, it was greeted with skepticism, even hostility. Consider the following from a 1990 article, "Some Scientists Doubt the Value of 'Genetic Fingerprint' Evidence," in the *New York Times*:

> Leading molecular biologists say a technique promoted by the nation's top law enforcement agency for identifying suspects in criminal trials through the analysis of genetic material is too unreliable to be used in court.
>
> The scientists say that for both theoretical and practical reasons the method, called DNA fingerprinting, cannot be counted on to decide with virtual certainty whether a person is guilty of a crime. Many say they are leery enough of the method that they would not allow their DNA fingerprints to be taken if they were innocent suspects in a criminal case.

Like many new, disruptive technologies, DNA testing went through the same cycle of techno-skepticism and dystopian fear before it became a trusted pillar of the justice system. Over that time, the technology improved, and DNA testing has only grown more reliable since. These days, the collection and use of DNA in criminal trials is standard practice. With good reason: the ability to examine DNA is one of those technological advancements that has meaningfully altered our justice system. Crimes that went unsolved for years suddenly had evidence, suspects, and convictions.

There are countless examples from organizations like the Innocence Project that testify to the power of DNA evidence to free the wrongfully convicted. DNA has also been instrumental in solving crimes that seemed to have no answer. In one of the more powerful examples, Gary Leon Ridgway was arrested in 2001 after DNA evidence linked him to the murders of four women. Known as the Green River Killer, it turned out that he wasn't just responsible for the slaying of those four women. He was behind forty-nine separate murders throughout the 1980s and '90s. The use of "unreliable" DNA is part of the reason the Green River Killer is behind bars today.

THERE'S ONE other important element to the DNA story: it remains today an imperfect system. Even though it's proven its value and has been used for decades, the collection and use of DNA still isn't a straightforward, perfect process. Crime labs have multi-thousand-case backlogs of cases in which DNA is waiting to be analyzed. Technology that speeds up the retrieval of DNA and makes using it easier hasn't been implemented in every crime lab in the country. And even though DNA has been commonplace in criminal justice for years, many professionals in the system often do not have the latest training and assistance to use it appropriately.

This final fact can prove pivotal in court. Many of us remember the 1995 O.J. Simpson trial as a prime instance of public skepticism over DNA evidence. But remember: Simpson's defense directed the jury's skepticism not at DNA evidence itself, but at the police officers and technicians who allegedly mishandled samples from the crime scene.

I mention the above issues with DNA as a way of making a broader point: the introduction, adoption, and use of new technologies in any field take time and patience, and even after their adoption, they are far from perfect tools.

A powerful example: in 1990 in Montgomery County, Texas, a man named Roy Criner was convicted and sentenced to ninety-nine years in prison for raping and murdering a sixteen-year-old girl. Even though witnesses gave vague and conflicting accounts of some of Criner's statements, Criner was put away. It wasn't until 1997 that DNA evidence established that Criner couldn't have been the rapist.

Even then, an appeals court was unpersuaded by the DNA evidence and Criner's conviction stood. A freelance writer examined the case and published an article, which drew enough attention that a new judge ordered additional DNA testing. That DNA testing finally led the courts to conclude that Criner couldn't have done the crime. Texas Governor George W. Bush pardoned Criner in 2000. Only after a decade of trials and prison time was an innocent man set free.

Note that three full years passed between DNA evidence that proved Criner couldn't have committed the crime and the ordering of *additional* DNA evidence that proved the very same thing. This reminds us that—no matter how effective a technology is—the work of criminal justice remains a fundamentally human endeavor. It's subject to delay, doubt, mistakes, misgivings, and apprehension. People in the field need to be taught and retaught how to use technology. But the presence of those hurdles shouldn't prevent us from experimenting, and as the DNA example proves, the use of technologies to help us get at the truth in criminal justice cases can save lives and advance the cause of justice by freeing the innocent and convicting the guilty.

The risk to society is that, for many of these emerging and unproven technologies, overreaction to the early bugs could be a death sentence from which there is no recovery. The world would be a very different place today if early skepticism about DNA had hardened into outright bans against it. Early reactions can be overreactions to perceived issues with technology rather than to actual technical problems. One of my goals with this discussion is to fight back against those premature reactions—to create a sense of injustice at the insufficiency of today's status quo in the hope that some promising technologies get a second or third chance to make a positive difference.

THINK ABOUT one technology that is both intriguing and troublesome: fMRI machines that purport to prove when someone is lying. Research on these technologies has been going on for about a decade, and several companies have begun the process of commercializing devices that claim to be able to separate fact from fiction. No Lie MRI and Cephos already have available applications of neuro-imaging technology; they say they can predict with better than 90 percent accuracy whether someone inside one of their machines is telling the truth.

But are *they* telling the truth about the efficacy of their technology? The theory behind these devices is simple: an fMRI machine can measure where blood goes in your brain. If, as these firms argue, it takes more effort to lie than to speak the truth, that would be reflected in blood flows—and, the theory goes, in brain scans.

What was once thought to be science fiction has moved closer and closer to reality. In 2007, no less an organization than the MacArthur Foundation—the same group that awards its annual Genius Grant—donated millions to investigate these intersections of law, technology, and neuroscience. Those MacArthur studies didn't fare well. Articles in numerous journals pointed out significant issues with using fMRI technology to illuminate truth or falsehood. The reason? It's currently impossible to infer a specific mental sequence just on the basis of blood flow in the brain. In other words, just because areas of the brain light up on a screen doesn't mean that a specific mental process—telling a tall tale, confessing the truth—is actually taking place.

Partially on the basis of those studies and others, the courts have held that fMRI results are inadmissible as evidence. For instance, in a 2006 murder case, *State v. Gary Smith,* Judge Eric Johnson of the Maryland Sixth Circuit ruled that "the use of fMRI to detect deception and verify truth in an individual's brain has not achieved general acceptance in the scientific community." Note that the court clearly left the door open for this type of technology once it achieves scientific validity.

As optimistic as I am about the potential of technological reform in our criminal justice system, I agree with the court's assessment. It probably is premature to introduce fMRI evidence into courtrooms, in part because fMRI is still an immature technology. That said, a projection of what this technology could mean for criminal justice is helpful, if only because technology can progress at a rapid pace and because those kinds of mental experiments can help us think about where to direct scientific effort and resources.

Imagine, for example, a defendant who—as happens in many cases— has a rock-solid and believable alibi, but no one to corroborate it. A great deal of other evidence in their case points in the direction of guilt, and an ambitious prosecutor is looking to sway a jury into voting for conviction. The defendant and his attorneys are firm: he wasn't there when the crime took place, so how could he possibly have been the one to commit it?

Enter the lie-detecting fMRI or its equivalent. The defendant volunteers to be placed under the fMRI machine and is asked a series of questions, similar to the ones used in lie detection today. Basic questions—"Is today Tuesday?"; "Is the sky blue?"—give way to questions directed at finding out what happened on the day the crime took place. As one additional data point in the defense's case, the neuro-imaging suggests that the defendant is telling the truth: the brain scans hint that he's being honest. Before fMRI is accepted as bulletproof, it may be accepted as one additional indicator amid other evidence, and like all that other evidence, it would be subject to differing levels of reliability. Taken together with the other evidence, should the fMRI lie detection results be allowed in court? Today, the answer is no, but it isn't too outlandish to imagine that, in the future, such a result would be useful for judges and juries who, ultimately, are making consequential decisions based on imperfect information.

That future may be a long way off, but as with many of the technologies we've discussed, the point isn't necessarily the specific technology or use case. The point is to reckon with the changes it will have on the criminal justice system, and, indeed, for the very idea of what justice is and what it means. If evidence in a case is no longer limited to witness testimony or the hard materials brought forth by a defense counsel or prosecutor, but instead begins to include what a defendant's brain might reveal about the truth, we need to upgrade how we talk and think about concepts like justice.

The idea of an fMRI machine that tells us what's inside someone's brain might seem a bit too far-fetched for some. So consider the technology that, today, gives us a wealth of new information about what actually happened on a given day, in a given time and place. I'm speaking, of course, about the tsunami of information that's available from smartphones.

Much of our personal information, our personal communications, and photos of our private moments reside now on our smartphones, and much of that evidence has been deemed admissible in a courtroom (on a case-by-case basis). The smartphone is opening a new legal frontier in the balance of privacy and security, even if we had never intended it to do so or even thought of the implications of that technology in the courtroom. In fact, one of the key issues in the case against Adnan Syed was whether or not cell phone towers could recreate a digital trail that would have put him at or near the murder site when it took place, or if he was many miles away.

When technology touches our systems of justice, it raises all kinds of complex and interesting questions—but in many cases, these are just "upgrades" on questions that society has been wrestling with since its origins. The Founding Fathers of the United States, for example, struggled to figure out what laws would be appropriate to the available technologies of their day. Under what circumstances, for instance, would an agent of the government be able to invade a person's home and search through their private properties? We created balances of power to maximize the ability of the government to protect the common interest while controlling the potential for abuse of individual rights. Those original discussions and debates cast a long shadow: even today, it's because of the Founding Fathers' concerns about government power that a police officer is required to obtain a warrant from the courts in order to search a person's residence, car, or smartphone (in most cases).

JUST AS we imagine the potential for the tech of the future, it is worth looking at the technologies in our justice system today that might need an upgrade. Consider, in that spirit, the performance ritual of a trial by jury. The jury trial is a means of accessing the truth to the best of our ability. And when it was first conceived, it represented an upgrade on the capricious decision-making of a king, warlord, or despot. Compared to alternatives—conviction without trial, trial by ordeal, or trial by combat—the jury trial was a merciful and meaningful step in the direction of justice.

But today, jury trials are an imperfect tool at best. Trial by jury is subject to all kinds of irrational pressures. A great deal of the outcome depends on emotion, performance, justice-as-theater. Lawyers know that showmanship and theatricality, often more than raw facts, can sway a jury and change the outcome of a trial. As the poet Robert Frost once put it, "A jury consists of twelve persons chosen to decide who has the better lawyer." It's why trial lawyers come up with clever quips—think of Johnnie Cochran in the O.J. Simpson trial, referring to the glove as evidence: "If it doesn't fit, you must acquit." This was as close as the early 1990s would come to a meme; that line and its resonance spoke to the fact that Cochran was trying to sway a jury, to plant a seed of doubt in their minds.

Just to be clear, I don't say this to belittle jurors—they do what most of us would do in their situation. Trial by jury is trial by twelve ordinary, imperfect human beings. But it is certainly worth asking, could we not improve that system? Are twelve randomly selected individuals who (usually) don't want to sit in a room and listen to the details of a case the best process we can come up with for determining whether someone is innocent or guilty of a crime?

Other societies have asked that same question and come to conclu-
sions that differ from our own. In 1969, South Africa banned jury trials
altogether. For decades before they instituted a formal ban, jury trials
had already been on the decline; people were electing to have judges,
and not juries, administer verdicts. Just like in the United States, ordinary people were often reluctant to serve on juries, and they would find
their own ways to skip service. Many in the justice system also pointed
out the obvious: that members of juries brought to the courtroom the
same prejudices they carried outside the courtroom. In fact, when
South Africa's *Abolition of Juries Act* was finally ratified in the late 1960s,
it was driven in part by the concern about racial prejudice among jury
members.

Legal professionals and scholars have long called into question the
value of the jury system. In fact, a 2007 Northwestern study found that
as many as one in eight juries makes the wrong decision. That's right:
in potentially one out of every eight cases, a guilty person will go free
or an innocent person will be convicted.

Critics of the jury system point out that jurors are often ill-equipped
to deal with very complex cases. Most jurors have no legal training and
no background in criminology. As often as not, people do what they can
to duck out of jury duty. If you serve, you often serve reluctantly and
with no knowledge of what you're judging.

But even if they are willing to participate, it's worth asking whether
twelve everyday people can be asked to litigate on matters that lawyers and judges go through years of professional training to even begin
to understand. Or, as law and psychology professor Peter van Koppen
put it in a 2009 popular essay on this subject: "A scientist has to make
inferences about states of affairs that cannot be observed directly,
inferring from evidence that can be observed. And that is precisely
what a jury has to do: make a decision about the guilt of the defendant
based on the evidence presented at trial. That is a scientific enterprise
that surpasses the intellectual aptitude of most laypersons who are
called to jury duty."

I would make the argument a bit differently: a jury trial is a type of
technology. It's a device that helps us accomplish an aim: the reliable
and fair administration of justice, or more simply, getting at the truth

of what happened. But just as with all technology, we can improve upon it. Once we have the ability to know the truth more reliably—to determine accuracy and falsehood with higher probabilities—we need to ask ourselves how long our current means of assessing justice are going to remain unquestioned and unaltered.

IN ONE DEEPLY important way, the justice system has already been reconfigured from the ground up. A decade ago, the primary way of determining what happened in a police encounter was by extensively interviewing witnesses, victims, police officers, and anyone else who was there when police arrived on a scene. Then a judge or jury would have to take that accumulated information and make a judgment about the truth.

Today, the truth is often available to us more simply and elegantly: it's recorded on video. The body cameras worn by police officers have fundamentally changed police work and the administration of justice. Few technologies have been more controversial in their development and deployment. A decade ago, the idea of police officers wearing cameras was perceived on both sides as invasive.

Early on, the most vocal resistance to uniform body cameras came from police officers themselves. Officers were worried that this was just another tool to monitor them, for bosses to harass the rank and file. They weren't even necessarily worried about the public. They were discomfited, as anyone would be, by the idea of being recorded all the time while on the job. If they cussed, would they be reprimanded? If they forgot to wear their uniform hats, would they be chastised by the higher-ups?

In the same way, the public response to body cams wasn't always a full embrace. When body cameras were first used, bloggers would write stories about how they were built "to let the police off the hook." There were concerns about body cameras being used for surveillance or to invade the privacy of people's homes. Was this just another tool for the state to spy on its own citizens?

This movement started, in earnest, with a tragedy, when Officer Darren Wilson shot Michael Brown in Ferguson, Missouri, in 2014. If you believe Officer Wilson, he was fully justified to fire his weapon in self-defense, as he was under a brutal attack from a larger, more powerful man. If you believe other accounts, Wilson shot and killed a man with his hands up who was attempting to surrender. (Subsequent investigations largely found that Brown was assaulting Wilson and disproved allegations that Brown had his hands up.) Following the incident, and the massive civil unrest that swept through communities across the country in its aftermath, Brown's family called for police officers to wear body cameras so that future families and communities could know the truth of what happens in police encounters.

Today, just a few years later, over 70 percent of the large city police departments in the United States issue body cameras to their officers. Along the way, there was legal resistance, union pressure, privacy issues, technological challenges, and more. And yet, today, this technology that was once thought to be strange and suspicious is considered normal and necessary, as much a standard part of every police officer's uniform as a badge. Body cameras—once loathed on all sides—are now acceptable. In fact, many skeptics now believe police officers should not be allowed to carry a lethal weapon without wearing a camera to provide oversight and accountability.

The story of Sergeant Brandon Davis shows the dramatic ways in which a piece of technology can change the outcome of a police encounter. One evening in 2009, he was on duty and responding to a domestic battery call. When he walked into the kitchen of the home, he saw a man, Eric Berry, with a gun in his hand.

"I told him to drop his weapon, twice," the officer later told the *New York Times*. But that wasn't quite the truth. In fact, a video of the incident, filmed on a body camera, showed that he had actually asked Berry to drop his weapon nine times. And it wasn't until Berry raised the weapon and started aiming at Sergeant Davis that Davis fired the shot that killed Berry.

Davis has watched the video many times since that night. And it was the video that saved his name: the shooting happened on a Wednesday, and by the following Monday, the investigation had been completed and he was able to continue on his job. The technology that rescued him from prosecution or months of uncertainty about his own future was an early version of wearable video. It captured, moment by painful moment, what happened that night, and it allowed the police officer, his department, prosecutors, and, eventually, a court room to see what Sergeant Davis saw in the kitchen: a criminal unwilling to put a weapon down, then pointing it at the officer, readying his aim, and preparing to shoot him.

Another powerful story involved Chief Jeff Halstead of the Fort Worth Police Department, and it showed how attitudes toward technology can change because of specific incidents. It was June 28, 2009, the fortieth anniversary of the Stonewall riots. On June 28, 1969, a police raid on the Stonewall Inn in New York's Greenwich Village turned violent. That raid was a seminal and scary moment for New York's LGBTQ community, but that fact wasn't top of mind to Chief Halstead's officers in Texas. On that June night in 2009, they were doing a routine sweep of the Rainbow Lounge, a local bar. It happened that the patrons of the bar were peacefully commemorating the Stonewall riots. "Our patrol officers were making bar sweeps, and at the end of their sweep, they decided to do a walk-through of the new gay bar in town," Chief Halstead told me. "When they went in they created an extreme amount of paranoia, shock, and resistance—and it turned into basically a bar raid by the police."

A photo of a police officer appearing to punch one of the bar patrons in the head went viral. It became a global headline, and national protesters descended on Fort Worth. Things became heated and the media dialed up the volume. Chief Halstead was facing pressure on both sides. The department was afraid he would summarily fire every officer involved in the raid, and the media thought he wasn't doing enough to take action against the officers, even though they didn't know what had actually happened in the bar. (Extensive interviews were conducted with the bar owners, patrons, and officers; the evidence file grew to be thirty-six inches thick. The review of the incident found the officers had used appropriate, restrained force, and that only one use-of-force incident required medical attention.)

Out of a tough situation came some important work: a community task force that worked to improve the relationship between the Fort Worth police and the LGBTQ community. A series of changes were made to the city's police department policies and ordinances. Twelve months after that incident, the chief was given a community leadership award from one of the organizations that had criticized him.

What it taught Halstead was the value of photographic evidence—and the danger if that evidence is wrong. In the case of the photo at the gay bar, the officer, it turned out, had not been using excessive force, but the situation was painted as such by the media and by activists. It made Chief Halstead want to figure out how to use evidence gathering and body cameras the right way so that situations of that kind could be avoided.

He didn't have buy-in from his fellow officers, at least at first. "There was resistance from the police officers' association," Halstead said. So he allowed the police officers' association executive board to write the policy about how discipline based on video evidence would be used during the first year. That one-year window was important—Halstead didn't try to reinvent the system overnight or force something on police officers that they were skeptical of. He worked with the police officers' union to make sure that they felt comfortable with the discipline that he would mete out as chief. "That was absolutely key to know that they're going to be okay wearing this technology and having every one of their contacts be public," Halstead said. "It was a generational and culture shift within the profession, and it was the right thing to do."

Even with the police officers' union's buy-in, Halstead's officers were often tough and vocal about how much they hated the idea of cameras watching every interaction they had in their jobs. So Halstead came up with clever strategies to promote buy-in. In a few instances, when an officer did something wrong, he would trade a slightly decreased disciplinary action in exchange for their wearing a camera for a full year and recording every interaction they had in uniform. Every officer subjected to this was willing to make the exchange, if begrudgingly. And Halstead took pride in the fact that "two of them, more than two years later, told me that their cameras had recorded evidence that saved their career from erroneous and false allegations of civil rights violations."

Over time, both the community and his officers bought into the value of access to information—the ability to deduce what actually happened at a given place and time without having to depend on the fogginess of memories, conjecture, or guesses. As Halstead put it, "We are, without a doubt, seeing exactly what was taking place prior to an officer's arrival, during the officer and multiple officers' arrival, and what happened afterwards. And I think that video is key because if we go back to a time when there's no video at all, it always remains a police-controlled headline." In other words, we can see right away whether an officer did, in fact, act aggressively. We can know, for certain, what a suspect was doing right before a violent incident took place.

One effect of this surveillance is that it speeds up a justice system that can be impossibly slow moving. It used to be that complex cases with a lot of evidence that wasn't digitized would drag on while paper documents were slowly moved from one lawyer to a judge back to another lawyer, and so on. Digital evidence means instant transmission.

Chief Halstead, who has been on all sides of these cases, put it like this: "At the same time that a prosecutorial agency receives the entire investigative file, it is literally duplicated and put into the hands of any defense attorney that is either going to be assigned or that will take on this case for their client . . . The fact that they have what they call an even playing field—both sides have complete case files with every single piece of digital evidence that existed from the start of the investigation, through conclusion, confession, and arrest—they both have a fair playing field to strategically advance their arguments before the judge. Judges appreciate this level of not just case transparency but expediency. Their backlogs are starting to decrease."

Prosecutors now admit that they are less likely to prosecute a case when there is no officer body camera video available. It's even become an expectation from juries, and they are suspicious, even stunned, when there's no body camera video present. And when it is available and used, the length of time it takes to render a judgment has gone down. That's a positive for both the innocent and the guilty, and it's a move in the right direction for a justice system in which both speed and efficacy could be improved.

IN A WORLD of imperfect justice, new technologies can get us just a little closer to the truth, to making our judgments about difficult situations based on facts, not stories; based on evidence, not guesswork. But even outside of the specific use cases for technologies, part of what we need to consider when thinking about the future is how technologies affect culture and customs.

Take, for example, the ubiquitous smartphone. The primary purpose of these devices when they were first launched was to make phone calls and access our messages. Could anyone have foreseen the profound cultural shift this technology caused? Could we ever have predicted that a phone could order us a car, lunch, new shoes, and the latest song we want to listen to, all seamlessly and in a matter of seconds? Could we ever have anticipated the unintended consequences, both good and bad, of a society in which every single person walks around with a photo-taking supercomputer in their pockets?

In the same way, body cameras have gone well beyond their intended purpose. Yes, they are still used to see what happened *after* a police encounter. But what's been surprising and revealing is that body cameras have had their own kind of deterrent effect: they've led to a reduction in both the use of force and citizen complaints about police.

In 2013, California's Rialto Police Department ran one of the earliest tests of body-worn cameras on their force and tested exactly this premise. The questions were simple: Will wearing personal video cameras decrease the number of citizens' complaints lodged against participating police officers? And will wearing personal video cameras decrease the prevalence of incidents involving police use of force?

Yes, and yes. During the two years in which this experiment was run, citizen complaints against police officers went down 90 percent and use of force went down 60 percent. A more recent 2016 study out of Cambridge University found that "use of body cameras see complaints against police virtually vanish." The study followed almost two thousand officers in both the United States and United Kingdom, and documented a 93 percent reduction in complaints against police officers, suggesting the cameras yield changes in behavior that calm potentially volatile encounters.

It's a bit uncomfortable as a police officer to know that you're being watched on the job, but it has the effect of changing your behavior, and perhaps unexpectedly, the behavior of the citizens you interact with. This might not seem like Earth-shattering stuff, except when you consider that few interventions in police work result in such a dramatic change in behavior. Police departments, like most organizations, tend toward certain behaviors, and changing those behaviors, like changing anything that human beings do, is notoriously difficult.

That's the upshot of new technology in criminal justice: the ability to change behaviors—subtly, imperceptibly, but effectively. The reason body cameras are important isn't just because they affect the lives of individual police officers and the citizens they are sworn to protect. It's because they are a useful case study in how technology can change the situations we take for granted.

WEARABLE CAMERAS are dramatically changing oversight of public safety officers, but the same technology could also open up innovative approaches to corrections and rehabilitation. In fact, as of this writing, there is a burgeoning field of offender monitoring, a cost-effective alternative to imprisonment. Most commercially available systems provide a GPS tracking ankle bracelet that allows for smarter parole systems and better punishments for offenders—something like an intermediate response that is more severe than a simple financial penalty and less severe than imprisonment. These are primitive systems, but as they evolve, I suspect they will change parole and post-prison administration as much as body cameras have changed police work.

Imagine, for instance, pairing a body camera with a GPS tracking bracelet for offender monitoring. Much as a body camera provides oversight of public safety interactions, the same device could serve as an accurate way to monitor the activities of ex-offenders or first-time offenders, at least up until the time when society has deemed that they no longer need to be monitored at all. One could argue that such an approach could be both far more cost-effective and more likely to avoid further escalation of criminal relationships than simply sending low-level offenders to prison.

As we've seen time and again, and as the research has confirmed, sending a first-time offender to prison can do more harm than good. It can immerse them deeper into the culture of gangs and violence. However, sentencing a first-time offender to wear a GPS tracking system paired with a body camera would likely serve as a gang-repellent system. A person wearing a camera that is being monitored by both AI and law enforcement personnel is unlikely to become involved in criminal activity, and the camera is likely to have a strong deterrent effect on bad actors who might otherwise associate with the offender.

For a few hundred dollars, it would be possible to establish a network of home security cameras paired with a wearable camera and tracking devices that could enable a "virtual incarceration model." Such a model could yield much improved outcomes at much lower costs than traditional incarceration approaches.

Does this idea seem strange to you? Or does the idea of scanning someone's brain to understand whether or not they did what they stand accused of doing feel odd? It probably does. Remember: we need to evaluate these technologies and other ideas we come up with not against a standard of "weirdness" but against the very real—and often troubling—outcomes of today. One study put the estimated annual cost of our current model of incarceration at a staggering one trillion dollars. Each year, countless lives are wasted in prisons that do nothing to rehabilitate people who have been convicted of a crime. The stakes of unjustly destroying a life through extended incarceration in a prison system are high. And what's worse, those prisons contain people who didn't commit the crime that a court said they did.

We need to ask ourselves whether invading personal space with cameras or fMRI technology is as strange or creepy as it seems when it's compared to a system that, today, takes away life and liberty and the hope of a future, and often does so in error. Few things could be more terrifying than being accused of a crime in which there are no witnesses other than the accuser. In that circumstance—a case with two people making a claim and no evidence to verify that claim—would an innocent accused person want state investigators to be able to scan their brain (and that of the counterparty) in order to ascertain the truth?

We can imagine they would—but only if the technology was highly reliable. That should be a mandate for the state. And we need to evaluate carefully at what point the reliability of the information would improve over the status quo, and at what point the information would be considered credible enough to be admissible as evidence in a courtroom.

Regardless of how close we are to science giving us the truth we're looking for, there is a very real human cost to allowing an imperfect status quo to carry on for years. Of course, I can't give the final, definitive answer on which technologies will change the work of policing, law enforcement, criminal justice, or soldiering. As someone who has operated in those worlds for almost my entire adult life, I know that it is difficult to predict where technological trends will go. Here's what I do know: often the thing that prevents us from making progress isn't the limits of the technologies that we have; it's the limits we place on our thinking about what those technologies can and ought to do. I hope you've seen that many of our limitations in criminal justice aren't limitations of resources or technology, but of imagination and ideas. If we learn anything from even our recent history, though, it's that ideas that can seem strange and uncertain can become a fundamental part of our lives.

Uncertainty, in a way, will always be part of the work of criminal justice. But the goal of the system as a whole is to do whatever it can to bring the uncertainty about what happened in a specific moment down as close as possible to zero. Justice is, to borrow a term from mathematics, asymptotic: it's a curve that moves in the direction of truth without ever hitting it.

Our system of justice is, in other words, a system for dealing with doubt. Evidence, witness testimony, the swearing in of key players in a case, rules of procedure—all of it is designed, in one way or another, to reduce the doubt we have that a court's ultimate ruling is just. But nothing can make that doubt disappear entirely. Seen in this light, technology can be a powerful tool for dialing down uncertainty by increasing the amount of information we have available to us. I believe that, much in the same way that DNA helped both to convict and exonerate, or body cameras helped to capture information about police incidents, the new era of criminal justice technologies will provide judges, juries, and courts with more information than they've had to date.

And that means that we might one day look back on *Making a Murderer* or *Serial* as fascinating, tragic historical documents—records of a time when our system of justice was far more plagued by doubt.

10

ALEXA,
CALL FOR HELP

SOMETIMES, THE NEWS gets a bit ahead of reality. You're scrolling through your social media feed, and you see the headline for a story that ought to be true—it looks true enough—and so you go on with your day assuming that it is. Odds are, you don't even click on it. You assume the story is true because it jibes with other true things that you assume about the world; it fits into your pattern of beliefs and expectations. And if it turns out later on that the story wasn't true, odds are that you'll never be aware of it. Corrections almost always get less exposure than the stories they correct.

Now, you might imagine that I'm about to launch into one of those lectures about "fake news" and the dangers of "alternative facts" that we hear so much of these days. But that's not where I'm going. Sometimes, the misreported stories that we accept without a second thought, without even clicking, are valuable and important for what they are. They don't tell us facts about the world. But they do tell us—or rather, our reaction to them tells us—facts about our expectations, facts that might not otherwise come to the surface. And so when we're fortunate enough to read both the misreported story and the correction, it's valuable to pause for a moment and ask what we've learned.

Let me give you a concrete example. Here's a news story you may have heard. In Tijeras, New Mexico, just east of Albuquerque, a twenty-eight-year-old man and his girlfriend, along with their daughter, were house-sitting for a neighbor. The man and his girlfriend got into a violent altercation. He assaulted her, and as the situation escalated, he pulled a handgun and threatened her with it. At some point, still holding the gun, he asked her, "Did you call the sheriffs?" She hadn't. But unbeknown to the abuser, the home where they were staying was equipped with an Amazon Alexa system, and fortunately for his girlfriend, Alexa's voice-recognition software is still working out some kinks. Alexa interpreted "Did you call the sheriffs?" as "Call the sheriffs." And so it did, placing a 911 call that brought the Bernalillo County Sheriff's Department to the door and possibly saving the woman's life.

As I said, you may already be familiar with that story. It went viral for a few days in the summer of 2017, and you're likely to have at least scrolled past the headline. If you read the story, you probably saw it all wrapped up with this quote from the county sheriff himself: "The unexpected use of this new technology to contact emergency services has possibly helped save a life. This amazing technology definitely helped save a mother and her child from a very violent situation."

Here's the correction: many outlets ran it just two days after the initial story, though with much less viral fanfare. Alexa could not, as of 2017, call 911. It could only call other Alexa-enabled devices, and only if the users of the devices on both sides gave permission. It could not sync up with a Bluetooth system to place any call at all to an outside number. It could certainly not respond to a command unless it was preceded by a "wake word," as in, "Alexa, what's the weather today?" And even as Alexa and its Google- and Apple-made competitors acquire the ability to place voice calls, that ability will likely not include 911 calls.

Wired magazine explains why: "According to Federal Communications Commission spokesman Mark Wigfield, providing 911 services means adhering to a host of technical regulations, everything from making sure all 911 calls route through the right call center, to making sure each one transmits the correct location of the caller. Additionally, devices that make 911 calls must also be able to receive incoming calls, so police can call back. Those hurdles currently prevent Google and Amazon from offering a direct emergency line."

In Tijeras, there was a 911 call—probably from the victim's cell phone. And there was an Alexa in the house. In the immediate confusion of the situation, the sheriff's department seems to have legitimately believed that the life-saving call came from Alexa, and those of us who read, clicked, and shared the story seem to have believed the same thing. Amazon issued a statement correcting the story days later, but while Amazon explained that Alexa did not, in fact, place the call, everyone involved took for granted that such a thing, in 2017, had become entirely plausible.

So what do we learn from this? To me, this is a case in which the news got slightly ahead of reality. The reaction to an initial bit of "fake news," from the sheriffs and then from the reading public, suggests that there's no longer anything especially surprising about a digital assistant taking steps to save a life. It's within our realm of expectations. The reasons it didn't happen are technical reasons, fixable reasons—but not conceptual reasons. Life-saving 911 calls from in-home artificial intelligence units are already part of our mental landscape.

The really interesting question, as *Wired* pointed out, isn't "Did Alexa make a 911 call?" but rather, "Should Alexa make a 911 call?" And even more interestingly, "Should Alexa make a 911 call on the strength of its own judgment, without our asking first?"

As you can probably guess, my answer to these questions leans toward yes, when the right safeguards are in place and the benefits to society outweigh the risks. I think we're at a remarkable point in our history, in which technological revolutions will continue to steadily drive down the incidence of killing and violence. That's true when it comes to law enforcement and the military. And I also believe it's true when it comes to the safety of our homes and our persons. Part of the work will be done by continued progress in technology. But the more important work will be conceptual, that is, it will lie in rethinking possibilities and priorities.

It doesn't require sci-fi thinking—just some extrapolations of what's already going on in the world. For instance, have you ever driven through a moderately upscale neighborhood and seen small signs planted near many driveways that say something like, "STOP! This home is protected by…" and then the name of a security company? You're looking at deterrence in action, and even if you don't give it more than a passing thought, would-be burglars certainly do. The sign tells burglars that, should they try to break a window or pick a lock, the police will be alerted.

Those signs, and the security systems that back them up, are already preventing violence: the violence of a home break-in, the possibility of a burglar shooting or being shot by a homeowner, the possibility of the police pulling up in the driveway with weapons drawn. The sign near the driveway is backed by the implied threat of the police, but because it's visible and well understood, it reduces the chance that the police will be called. That's deterrence: like the famous "dog that did not bark" from the Sherlock Holmes story, we see it in crimes that do not occur and break-ins that are not even attempted.

Now, I also understand the skeptical view: home security systems don't actually reduce crime and violence, they just channel it elsewhere. By now, burglars are smart enough to anticipate the presence of home security systems, and they know to go where those systems aren't. It's reasonable to expect that home security systems don't stop burglary so much as push it slightly down the socioeconomic scale, into homes that are well-off enough to have something worth stealing but not well-off enough to afford expensive security measures.

The good news is that the cost of home security electronics, as of electronics in general, is steadily coming down. In chapter 6, I introduced Moore's Law, which observes that we can fit twice the number of transistors onto a silicon chip roughly every two years. The same forces that continue to drive the spread of small, cheap consumer electronics are also driving the spread of affordable home security.

As of this writing, you can buy an Amazon Cloud Cam, compatible with Alexa, and have it shipped to your home in two days for $120. It notifies your Alexa when it detects suspicious movement, it allows you to store and share twenty-four hours of footage, and it lets you watch the outside of your home on your Alexa monitor in night vision. And that's an expensive model! If you're willing to go off-brand, you can buy a two-pack of Wi-Fi–enabled home security cameras for $57 or a four-pack for $100. Or if you're a bit more upscale, a Nest camera from Google can beam suspicious activity to an app on your phone for $350.

Of course, the point of these cameras is that, ideally, they'll never have to be used. Burglars and other criminals see them and keep driving. As plummeting costs lead home security systems to be more widely adopted—as they become as standard as locks on doors—burglars will have to keep driving further and further to find targets.

That's a linear model of progress—the more homes with smart cameras, the fewer targets for property crime. But in fact, it may be more reasonable to think of this progress as exponential, due to network effects. That means that each camera gains in deterrent power as other cameras come on line. On a simple level, it means that a burglar is probably less likely to target your home if both of your neighbors also have security systems, and less likely to bother even casing your neighborhood if a majority of homes are similarly equipped.

On a more complicated level, we can imagine those cameras collecting valuable data and sharing it over security networks: about the times and conditions in which burglaries are most likely, about the license plate numbers of cars that have been spotted at the scene of previous break-ins, about potential blind spots or routes of entry. We could even imagine those cameras using facial recognition technology to identify known burglary suspects. All of these features become more valuable as more homes participate; that's the network effect. None of these features are foolproof, but together, they drive up homeowners' deterrent power and also drive up the costs of crime for would-be criminals.

Of course, a great deal of violence in and around homes doesn't obey that kind of cost-benefit logic. In the Tijeras case, just as in so many incidents of domestic and intimate partner violence, violence can't be deterred by a security camera, because the perpetrator is already inside the house.

WHICH RAISES another question: What active measures can people take in the event of facing a dangerous person, whether in a home invasion, a parking lot confrontation, or any other locale? The dangerous assailant is the primary use case for people to arm themselves with weapons of all types. Historically, the option most readily available to the public (i.e., firearms) have also carried the weighty side-effect of taking a human life in the process of protecting yourself.

There are powerful statistics on both sides of the argument around whether a gun is a good choice for self-defense. Gun-control advocates will tell you that fatal accidents, suicides, and homicides far outnumber legitimate killings in self-defense. Gun rights advocates will tell you that firearms deter millions of crimes each year. It's not my place to try to determine who is correct. However, I believe the conventional thinking that "the only way to stop a bad guy with a gun is a good guy with a gun" will change as non-lethal capabilities grow stronger.

That brings us back to Alexa (and its counterparts from other tech companies). To me, it seems a no-brainer that we can and should overcome the regulatory hurdles that prevent an in-home assistant from calling the police on demand, once the accuracy of the decision process to do so clears the right hurdles.

Earlier I said that Alexa can't call 911 for you. However, people who decide they do want Alexa to be able to call the police can add it as a skill using a third-party private personal safety technology company called Noonlight. Government 911 centers have to adhere to a strict set of regulations and requirements, so they can't adapt very quickly to new technologies, like Alexa calling in on your behalf. The team at Noonlight took a different approach: they created a private dispatch center that interfaces with Alexa or any other connected device. The Noonlight dispatch system is unencumbered by 911 system requirements and regulations, which means it can move at comparative lightspeed.

Add the Noonlight skill to Alexa, and you can now directly use Alexa to connect to emergency services through Noonlight's twenty-four-hour dispatch center. It has all of the technology interfaces to accept location and other data from your smartphone, Alexa, smart watch, or other smart devices. It has built-out data registration tools to prepopulate your records of fixed device locations, and it has all that data at hand when an emergency strikes. Of course, Noonlight can't directly send a police officer (yet), but it can calmly call your local 911 center on your behalf into a three-way call in which the Noonlight dispatcher can share key information that the police need, which might be a challenge for you to share directly if you happen to be running for your life.

I think Noonlight is on to something. Most of us don't want the police to be able to track our cell phones or listen in through Alexa in our kitchens. However, many of us would like the ability to set up our devices to support us in a time of need, notifying our parents, kids, or the police as appropriate—but only when we say it's appropriate.

I don't think most people would voluntarily sign up at their local police department to set up a bunch of tracking information about themselves and grant direct access to their audio and video devices. But over 1.5 million people have done exactly that with Noonlight. It seems that the existence of a trusted intermediary technology partner— not Alexa itself, but a service built atop Alexa—has solved the privacy threshold for many people to connect to their local police.

I should emphasize the importance of the word "trusted" in the previous sentence. Many people do not have the time or expertise to perform security and privacy audits to ensure that third-party private companies can be trusted. By adding a layer of control and privacy between your devices and the police, Noonlight has found one way to make Alexa calling for help not just acceptable but also desirable. It has solved the privacy issue by giving clear control to the user to configure exactly when the system calls for help. And it has solved the accuracy issue by inserting a human operator in the loop before things escalate to the police.

This is an important area for continued governmental and independent standards to provide oversight to help consumers avoid giving away their privacy to untrustworthy entities. Government oversight is advisable, of course, but I believe that private technology companies like Noonlight will have a critical role to play going forward in providing oversight of their own.

PERHAPS A more interesting case is whether Alexa should summon help without our explicit request. Say when its speakers pick up a phrase like "I'll kill you," or when its camera detects an unrecognized firearm being brandished in an aggressive manner. I could imagine a home safety package of this kind included in Alexa's default settings. Users with stringent privacy concerns would be free to opt out—say, after an authorized adult in the house enters their password—but I doubt that many will.

Wouldn't abusers simply opt out of a system like this, or pressure their partners to do so? No doubt, some would, and in those cases, we'd still be in the current status quo. But I'm willing to bet that a surprising number would not. Here's why.

I don't like the phrase "crime of passion" when it's applied to something like domestic violence. It can give horrific violence a kind of romantic, almost understandable air. Like all euphemisms, it's halfway to excusing the thing it claims to describe. That said, I do think the phrase can remind us of something. Unlike crimes that are conceived beforehand and follow a rough cost-benefit logic, domestic violence is often unplanned. Most abusers don't think of themselves as abusers. You can tell from the sorts of bogus excuses they rely on when confronted with their actions: "I wasn't in my right mind"; "That wasn't me"; "I'm really a nice guy." You don't have to accept the truth of those excuses to recognize the gap they open up between what abusers do and how abusers perceive themselves.

An Alexa home safety package is a way of exploiting that gap. In a calm moment, in which an abusive partner is acting like the "nice guy" he imagines himself to be, it's easy to opt in to the home safety package or to keep the default setting in place. And when he comes home drunk and starts threatening violence, the system is already there to protect his partner. In this way, the system I'm describing is a special kind of violence deterrent: a precommitment system.

A precommitment is a way of binding our future selves. A classic example is handing your friend your car keys when you arrive at a party. Your sober self knows that your drunk self shouldn't be driving, and when your drunk self shows up a few hours later to demand the keys, your friend knows to refuse. I could imagine Alexa operating in a similar way, kicking in to stop abusers when they are at their worst.

AS FAR-FETCHED as all of that might seem, it's important to keep in mind that technology is already playing a role in reducing violent crime. Over the past two-plus decades, violent crime in the United States has plummeted across the board (see figure 5).

FIGURE 5: REPORTED VIOLENT CRIME RATE IN THE U.S. FROM 1990 TO 2017

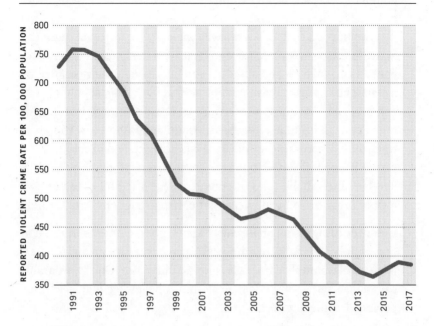

The reasons behind the Great American Crime Decline are much debated. They range from the phase-out of lead paint to the adoption of sophisticated crime statistics software, like the NYPD's CompStat system. But a share of the credit is due to the humble credit card. There's little sense in mugging or robbing a pedestrian who is likely carrying very little cash and who can cancel their cards instantly. The credit card isn't a flashy crime-prevention technology—few of us think of it as a crime-prevention technology at all—but it's made a measurable difference. Here are a few other small-scale technological interventions that can continue to drive down the violent crime rate:

- **RFID chips:** "Tracking chips" are everywhere from contactless entry cards, to E-ZPass toll readers, to library books. They can be as small as a grain of rice, and you can buy one online for just a dollar. As small, cheap tracking devices become more widespread, they can act as another deterrent to violent crime and theft of valuables by tracking stolen items and making it easier for law enforcement to apprehend thieves.

- **Public safety apps:** At Arizona State University, students are equipped with the LiveSafe app on their smartphones. It provides easy access to police and emergency services, but its functionality goes beyond that. Students walking across campus at night can activate the SafeWalk feature to chat with designated emergency contacts in real time and allow the contacts to view their progress on a map. The app also allows students to find their nearest public safety call box and to request a safety escort or ride home. Even if users aren't college students, they can use similar apps to allow friends to track their progress if they have to walk at night, and to contact local police if needed. The Noonlight service, described above, offers similar functionality and augments it with the ability to connect a wide array of smart devices that can also trigger alerts for help.

- **Life recording:** We usually hear about body cameras in the context of police work, but with the growth in live-streaming, the next frontier in wearable cameras is likely civilian use. Cheap, wearable cameras, whether embedded in clothing or eyewear, will be another powerful crime deterrent.

Now, in all of these cases—from in-home security systems to wearable cameras—we're likely to meet a familiar objection: privacy. We're likely, when discussing these technologies, to hear at some point this famous line from Benjamin Franklin: "Those who would give up essential Liberty, to purchase a little temporary Safety, deserve neither Liberty nor Safety." Perhaps we'll hear the word "Privacy" substituted for "Liberty," but the thrust is the same: trading something essential for security is a coward's compromise.

It's well said—Franklin had a way with words—but it's not a knockdown argument. For one thing, look at the qualifying words: "essential" Liberty (or Privacy) and "temporary" Safety. If the growth of social media over the past decade has shown us anything, it's that the vast majority of people are willing to compromise bits of their essential privacy, in part because they don't necessarily believe that anything "essential" is being sacrificed, or because their definition of privacy has changed in an era of ubiquitous smartphones and social media use. For instance, they're willing to let Alexa in their homes, because they don't see it as compromising their privacy in an essential way. Similarly, we're not talking about a "temporary" gain in safety, something like repelling an invasion by the British. Rather, we're talking about large-scale, widespread gains in the security of everyday life—the kinds of gains we saw in the Great American Crime Decline, which changed the face of our cities in the span of a generation.

But let's set semantics aside. The deeper premise of the Franklin-style argument is that it's dangerous to trade some of one right for more of another right. However, we trade off rights against each other all the time. No rights are absolute in practice: we balance them, negotiate them, and calibrate them to create the kind of society we want. I imagine you agree that there's no right to shout "Fire!" in a crowded theater. Well, if you do, then you also agree that the right to free speech can be traded, in some cases, for the right to public safety. The only real question, then, is what sorts of trades we should decide to make. Privacy is no different.

Franklin's line is pithy, but something said by his contemporary, the English philosopher Jeremy Bentham, is closer to the truth. Here are Bentham's words:

> What is the language of reason and plain sense upon this same subject [of rights]? That in proportion as it is right or proper, i.e. advantageous to the society in question, that this or that right—a right to this or that effect—should be established and maintained, in that same proportion it is wrong that it should be abrogated: but that as there is no right, which ought not to be maintained so long as it is upon the whole advantageous to the society that it should be maintained, so there is no right which, when the abolition of it is advantageous to society, should not be abolished.

That's a mouthful, but it's easy enough to boil down. We establish rights because they are useful. They are advantageous to us and to the societies we live in. And it's up to us to define what we mean by useful and advantageous. When a right serves our purposes, we should keep it. When it stops serving our purposes, we should abolish it. When it serves some of our purposes but gets in the way of others, we should scale it back.

Remember, it's not an oppressive government that is scaling back our experience of the right to privacy—we ourselves are doing it, in free, individual decisions, every day. Whatever we say, this is how the vast majority of us act: we value our privacy so little that we give it away to Facebook for free, in exchange for the ability to share cat videos. Those who dissent from that bargain don't have to take up arms; they don't even have to go off the grid and live in the woods. They just have to deactivate their accounts. That most of us do not, when the choice is sitting right in front of us every day, tells us the real value we place on privacy compared to our need for social connection.

But let's say that even those of us who make the choice to give away our privacy in those contexts worry about the infringement on our privacy when that data is being filtered through a police department, or a government agency, or even a third-party like Noonlight, whose objectives are not to connect us with friends or share photos of grandchildren. I think it's worth taking those legitimate privacy concerns seriously. I think there's also a way to accomplish those aims by building better technology and by carefully designing the controls that govern its use. It isn't a stretch to imagine a device AI that could evaluate what audio and video is worth keeping or sharing. You could imagine in-home cameras as observers, not as information-storage devices tracking your every move. Or imagine that same information being stored, but encrypted in a way that could only be accessed with a search warrant—in other words, only when it's ultimately needed.

I would argue that we shouldn't throw out the idea of technology because we haven't answered every question about privacy yet—urgent and important though those questions are. None of this should be an excuse when the companies we trust to protect our privacy let us down. The fact that we trust private companies more than our own government with our most private data conveys an important burden upon those companies to be transparent and careful with how they use our data. They should clearly and explicitly request our permission before using or sharing that data on our behalf. And there is a clear role for government to help set the rules by which industry must operate. We may trust Google more with our data than the National Security Agency, but we must rely upon the regulatory powers of the state to help ensure that private companies abide by a clear, transparent set of guidelines around how that data is protected.

And if we would trade away some privacy for cat videos, why wouldn't we trade it for something infinitely more precious—freedom from violence, safety in our own homes and schools, peace of mind? As I'll discuss next, we rarely know how precious that sense of safety is until it's taken away.

SCHOOL SAFETY

CLICK.

The sound of the magazine locking into place in his AK-47 didn't seem as impressive in real life as it had in *Call of Duty*. But the sense of power in holding a fully loaded assault rifle was intoxicating.

Randolph Jones hadn't slept in two nights, since the evening Judy had told him that they were through. Although they'd only been dating for a few weeks, it had felt like this time it was going to work. But, just like the others, she failed to understand him, failed to appreciate what he was capable of.

He'd stopped taking his meds that night. They clouded his thinking and he needed to think very clearly right now. Gaming had always been his retreat, a chance to melt away into a world where his skill set him apart, where he excelled. But this time, as he played through the night, day, and night, he realized it was time to demonstrate his power in the real world. Time to get the respect he deserved.

Judy worked at the Sunset Daycare Center. It seemed fitting, Randolph thought, that this would surely make the news worldwide. He wouldn't just run up the score today; he would be forever remembered. He'd be infamous. If they thought an elementary school was newsworthy, just wait.

Fifteen minutes later, he pulled up in front of the daycare and hesitated briefly before parking in the handicapped spot by the front door. "Fuck it," he thought. "I wonder if they'll bother to mail the parking ticket."

He threw the door open, feeling excited as he walked past the front desk and down the hall. The receptionist's mouth was agape, her eyes fixated on the weapon in his right hand. Or maybe at his swagger as he blew past her.

Judy's room was the first one on the right. He remembered that from last week when he had picked her up after work. The look on her face was quite different this time as he burst into the room.

Fear. Sheer, unadulterated fear.

For dramatic effect, he started by shooting at the windows near her. He had always wondered what it would be like to watch the sheets of glass shatter in cascades, just like in video games. The modifications he had made to the weapon functioned as expected, delivering a continuous stream of bullets in full automatic mode as he held the trigger down. He heard the rat-a-tat-tat of the weapon firing, the sound of glass shattering, and a growing chorus of high-pitched screams as everyone realized the power he wielded. The world would now experience his hate—and fear him for it.

Twenty feet away from the muzzle of his weapon was a circular device on the ceiling, resembling an oversized smoke detector. Milliseconds after the first round exited the barrel, the first sound waves passed over the array of microphones on the bottom surface of the white cylinder.

As confusing and terrifying as these bursts of sound were to the people in the room below, they were merely data to the microprocessor monitoring the activity from the microphone array. The AI algorithm was designed to constantly monitor for potential firearm discharge sounds, not all that different from the iPhones of millions of people around the world that are awakened by the "Hey Siri" sound pattern. In this case, the sound pattern fit the template of the AI algorithm, which immediately recognized a 99.8 percent probable firearm discharge. The algorithm quickly calculated the direction based upon the relative time of arrival of the first sound wave to each of the four microphones. Combined with the panic alert signal that had been activated at the front desk fifteen seconds earlier, the system escalated the probability of a firearm discharge event to be in excess of 99.9 percent.

Five hundred milliseconds passed as the system fully activated, firing the release mechanism that set the white cylindrical cover free and falling toward the ground. Another quarter second passed before the small quad-copter drone dropped free and began spinning its rotors, first to stop the free fall of the drone itself, then to orient toward the direction of the sound. The computer vision algorithm on board recognized bright flashes of light which, in combination with the continued sound waves hitting its microphone array, were identified as 99-percent-plus probable muzzle flash. Four independent events were combined: the human activated panic signal, microphone firearm discharge detection, camera detection of visual patterns of muzzle flashes at the same discharge intervals as the sound bursts, and now visual identification of a device determined with 90-percent-plus probability to be a firearm located adjacent to the muzzle flashes. Human confirmation through the building's panic system was used as authorization to engage and incapacitate any person holding a probable weapon.

Randolph had expected the sounds of shattering glass and screams, but this new sound was unexpected. A high-pitched buzzing caused him to look away from the glass and toward two objects that appeared to be falling from the ceiling. One object accelerated toward the ground, landing with a thud. The other behaved more strangely, slowing to a hover in mid-air.

"Drone!" Randolph's brain registered. But his was the slower processor. By now, the camera system aboard the drone had detected Randolph's face in addition to his weapon and the muzzle flashes. Tracing the outline of his body was fairly straightforward, by combining the main targeting camera feed with a laser depth sensor and muzzle flashes. Based on the relative position of the human holding the weapon, the drone selected two locations and fired one small dart at each. The darts flew toward their mark, tiny wires unspooling along the flight path back to the power supply in the drone. These wires would soon carry electrical impulses designed to paralyze the human nervous system. Only 1.75 seconds after the first sound wave had hit the microphone, the darts found their mark.

As Randolph began to swing his weapon toward this new and unexpected threat, he felt a surge of energy cascading through his body. His right arm went rigid, as did the length of the right side of his body. The sensation of invincibility evaporated, replaced by utter helplessness as he watched his weapon fall away from his hands. He watched his right wrist curl into the awkward pose associated with palsy, and he began what felt to him like a long descent to the ground. One second ago he was indomitable, hurling his vengeance at will. Now, he was trapped inside his own body. This was not how it was supposed to end.

He could hear the screams and pandemonium around him, and sensed people running everywhere. But he could not make sense of any of it; he was inwardly focused, frustrated at his sudden paralysis. Over time, the screaming of children subsided and the room was empty, but the surging energy and immobility continued.

Then it stopped, as suddenly as it had started.

About eight miles away, at the real-time crime center, an operator monitoring the CCTV from the daycare center issued a command over an integrated speaker in the room:

"This is the Flagstaff Police Department. Do not move or you will be shocked again. Repeat, do not move." If Randolph reached for the gun, tried to get up, or took any similar action, the remote command to reactivate would be transmitted.

Randolph took a halting, deep breath, and began to cry. Four long minutes passed until two officers entered the room and secured the weapon. The sound of handcuffs ratcheting closed was the last sound Randolph had expected today.

THAT SCENE, like the futuristic war in Raqqa, imagines what a violent incident inside a school might look like in the future. Today, of course, the world looks very different. The signs inside schools speak to the danger our children live with every day. Often, they're hidden at the back of a parking lot or near a rear entrance, but increasingly they're a feature of the landscape. "If you SEE something, SAY something," they read. "Report any suspicious activities or threats against this facility to local law enforcement."

The signs I'm thinking of aren't just in front of government offices, crowded subway stations, or other places we may think of as typical terrorist targets. They're in front of schools, where they're becoming as regular a sight as drop-off lines, soccer fields, and baseball diamonds.

This is what it means to send a child to school in the United States today: to accept that the building where you drop them off every morning might become a target for terror. To accept that they'll grow up practicing "shelter in place" drills, just as an earlier generation learned to "duck and cover" in case of an impending nuclear war. To accept that ordinary sounds—a school bus engine backfiring, a clank in the building's heating vents—will always sound different to them than they did to you.

Each news report of a school shooting provokes outrage, grief, and fear. Those emotions, whatever our political convictions, are rooted in a shared concern for our children and the security of their childhoods, and a shared sense that something is deeply wrong when children can't learn without fearing for their lives. Most parents I know would do anything—anything—to make that fear go away. That's only human.

But that doesn't mean that those of us who are affected by school shootings—which, in a country connected by mass media, means all of us to a greater or lesser extent—can't also be smart about what it takes to stop them. We need outrage, passion, and commitment; we also need clear eyes, ingenuity, and a willingness to think outside the box.

Before I dive into what has become an incredibly contentious debate, I want to stress one point: what I see as the more promising solutions—addressing school shootings through more innovative technology and smarter policing—aren't mutually exclusive with other proposals. Whether you think that more armed teachers would mean safer schools, or whether you think the answer lies in cracking down on gun ownership in general, I think we can still find some common ground. And that, in itself, is something to work toward.

THE MOST controversial proposal on the table to deal with this problem is the suggestion that we ought to equip every teacher in the country with a firearm so that they can protect their students in the event of a mass shooting. At the outset, I want to point out that I'm not going to engage in an argument over the right to bear arms or the Second Amendment. You can believe that the personal right to bear arms is protected by the US Constitution without assuming that it makes sense for teachers to carry firearms as part of their work responsibilities. We need to deal with the proposal to arm teachers on its own merits, not as a part of a long-running constitutional debate.

Here's my candid view: responding to the threat of school shootings by equipping teachers with guns carries a risk of turning schools into even more violent and fearful places. The idea of arming teachers is one approach using pre-existing technologies. But shouldn't we think more creatively about what it would actually take to make this a less violent society?

The idea of a gun-wielding teacher protecting students sounds appealing in theory, but the facts are stacked against its efficacy. I return, as I often do, to the experiences of police officers and their guns. Cops have to be comfortable handling a gun. They have to spend hours every year training in the safe and effective use of lethal weapons. And not least, they have to be psychologically ready to pull the trigger—ready to actually take a life when the situation calls for it, and ready to bear the consequences of making a bad choice in a split second, either by acting wrongly or by failing to act.

How many teachers have this training and qualities? No more than a small fraction, surely. In part, that's because police work and teaching attract different personality types. But more importantly, police work and teaching create different personality types, by exposing cops and teachers to widely different training. If we wanted the more than three million public school teachers in the United States to think and act like cops—to be comfortable with handling guns, to be willing and able to kill when an emergency called for it—we would have to train them like cops. Sure, that might be possible. But every hour we spent training a teacher to handle a gun would be an hour away from that teacher mastering their subject matter, designing lessons and grading papers, and mentoring students.

Imagine, though, that we've decided to invest in that training, and that we now have a nation of armed teachers, the quintessential "good guys with a gun" standing guard over our classrooms. Could we expect them to prevent school shootings? The data suggests that we shouldn't bet on it. Now, it's certainly true that armed civilians do prevent crime. Preventing crime takes guts, awareness, and extensive training—and if you've put those qualities to use to protect your community or your home, you have my respect. But while guns can be effective for self-defense, they are also very prone to accidental use or to catastrophic use in a moment of anger.

Christopher Ingraham, a reporter for the *Washington Post*, studied the FBI's nationwide data on shootings and found that these negative uses outweigh the positive ones. For every "justifiable gun homicide," a case where a gun killed a criminal in self-defense, "there were 34 criminal gun homicides, 78 gun suicides, and 2 accidental gun deaths." The same gun that is intended to protect a classroom from a mass shooter can be used in anger, discharge by accident, find the wrong target, or even be used for suicide.

On the other side of the debate there are studies, such as one conducted by the National Academies' Institute of Medicine and National Research Council, that find guns do have a significant deterrent effect. The report noted that "defensive use of guns by crime victims is a common occurrence" and "almost all national survey estimates indicate that defensive gun uses by victims are at least as common as offensive uses by criminals, with estimates of annual uses ranging from about 500,000 to more than 3 million, in the context of about 300,000 violent crimes involving firearms in 2008."

So you can understand why some people want to arm teachers: they want them to be able to defend children in the classroom. Proposals to arm teachers would train them in the use of firearms, so we would expect those statistics to look better for a group of trained professionals than for a random sample of civilians. And yet even the best-trained law enforcement officers struggle in mass shooting situations. Sheldon Greenberg, a Johns Hopkins professor and a former police officer, has spent nearly two decades studying school shootings and other mass shootings, along with law enforcement officers' responses. His work found that "despite police officers' extensive training and familiarity with high-risk and life-threatening events, the evidence shows that they do not shoot accurately in a crisis encounter."

Greenberg held roundtables with law enforcement officials in the aftermath of the 2012 Sandy Hook Elementary School shootings. He found that among the biggest opponents to arming teachers are the very same cops who would have to respond to a school shooting. Police officers had three concerns: 1) even armed teachers might not be anywhere near an active shooter; 2) training teachers in the use of firearms risks being treated as a one-time, check-the-box exercise rather than genuine in-depth education in how to use a deadly weapon; and 3) it's entirely possible that in the heat of the moment, a teacher might confuse a plainclothes police officer wielding a weapon as the shooter and inadvertently fire on the police officer.

That is, of course, if the teachers are able to shoot at all. Tragically, in the February 2018 school shooting in Parkland, Florida, the police officer assigned to the school failed to enter the building during the shooting; his defense was that he did not know there was an active shooter and had received reports of "firecrackers" being set off. The evidence gathered from police responses to ongoing mass shootings, Greenberg concludes, means that putting teachers into such a crisis would be a "crapshoot."

On top of all that, remember that teachers are human beings, too. Even if guns in classrooms are safely secured—and I'd prefer that they were put in biometric safes, to which only the teachers' thumbprints would grant access—those guns could still be compromised in a heated moment. Would the odds of that happening exceed the odds of a teacher using a gun to stop a shooting? It's hard to say. But it's reasonable to ask whether increasing the number of deadly weapons in schools will actually make schools less fearful places.

One other important factor to consider: the risk of accidental discharges. According to the NYPD *Annual Firearms Discharge Report*, there were eighteen unintentional discharges of firearms involving NYPD officers in 2014. Of those, sixteen were in non-adversarial situations, including two during loading/unloading and fourteen during normal handling of the weapon. These incidents resulted in four injuries and one death. The NYPD has about 38,000 officers. Accordingly, the rate of accidental discharges is 1 per 2,100 officers.

According to the National Center for Education Statistics, the US public education system employs 3.2 million full-time teachers as of 2018. If we make the simplifying assumption that teachers could be made as proficient with a gun as police officers, and we apply the same rate of 1 accidental discharge per 2,100 people per year, then we could expect approximately 1,500 accidental firearm discharges by teachers each year in public schools. Maybe the number would be lower, because teachers might handle the firearms less frequently. Maybe the number would be higher, as they would have less proficiency than trained police officers.

A 2018 report from CNN found that there have been 288 school shootings in the United States since 2009. This equates to roughly twenty-nine school shootings per year. If the estimate of 1,500 accidental discharges from teachers' guns turned out to be accurate, this might be too high a price to pay to address twenty-nine school shootings that might be prevented, deterred, or interrupted.

ON THE other side of this debate are gun-control advocates. Rather than arm teachers, they say, let's get rid of the guns that are the central source of the program.

Take the example of Australia, which is often held up as a beacon by American gun-control advocates. In 1996, Australia suffered the kind of mass shooting that has become sadly routine in the United States— the Port Arthur massacre, in which thirty-five people were killed and twenty-three wounded. In the aftermath, Australia passed sweeping gun-control legislation, restricting the private ownership of most firearms. Most notably, the legislation included a mandatory "buyback" program in which 643,000 guns were handed in to the national government.

Australia's gun-confiscation program was successful in bringing down the gun homicide rate and the incidence of mass shootings. But it's doubtful that the success of that program can be replicated in the United States. Consider again the degree to which the United States outpaces all other countries—not just developed countries, all countries—in gun ownership per capita. It's shown in chart form in figure 6 (overleaf).

That's 88.8 guns per 100 people, or nearly a gun per person (counting children) in the entire country. Our nearest competitor, Yemen, has just 54.8, or a bit over half our guns per capita. All told, Americans own about 42 percent of all privately held firearms in the world.

FIGURE 6: COUNTRIES WITH MOST GUNS PER CAPITA

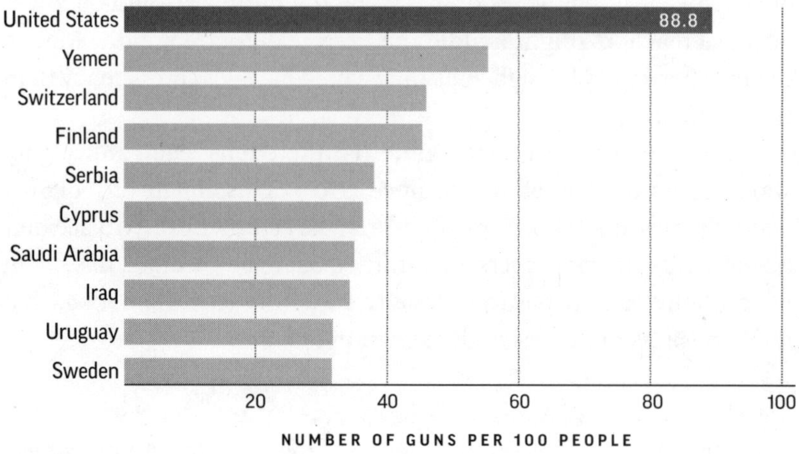

NUMBER OF GUNS PER 100 PEOPLE

Perhaps the United States could pass gun-control laws and impose stiffer penalties for breaking them, but at the end of the day, we'd still be the country with more guns than adults. We'd also be a country in which gun-control laws have, at best, a questionable track record in preventing gun violence. After all, the first modern gun laws in the United States were passed in 1965. Since 1968, there have been an estimated over 1.5 million deaths from firearms—with the number increasing each decade. The US legislative process has not yet proven capable of fixing the problem, a stalemate that shows no signs of changing. Like many, I'm pessimistic about this country's willingness to implement the amount of gun control that could make a measurable difference.

If the United States were to attempt an Australia-style program of buybacks and confiscations, it would be attempting it in a country with a far, far higher level of gun ownership than Australia's. Like it or not, the United States has a deeply held culture of gun ownership, one enshrined in its Constitution in a way that it is not in comparable states like Australia or the United Kingdom (which passed its own restrictive gun laws after a 1996 school shooting). A significant percentage of the American population believe that their right to own a gun is as sacred a right as any other, and they would forcibly resist any attempts by government to remove their guns. In a political culture in which one side's constant warning is "The government is coming to take our guns," imagine the fallout if the other side responded, "Yes, the government is coming to take your guns."

None of that means we can't optimize legal frameworks to reduce some of the risk. Less radical proposals—better background checks, better-enforced waiting periods, restrictions on gun ownership for those with domestic violence convictions, restraining orders, or mental health issues—might have potential to curtail gun violence. But barring an unprecedented shift in this country's gun culture, the fundamentals of the problem won't change anytime soon.

That's why I choose to place my hope in addressing the problem with innovation and better technology. Again, considering alternative approaches doesn't require you to give up your stance on guns, whether your views are closer to those of the NRA's Wayne LaPierre or Parkland student and gun-control activist David Hogg. It just requires you to open your mind to the possibility that we can make significant progress without resolving this country's long debate on guns once and for all.

SO IF more guns aren't the answer, and fewer guns aren't a plausible option, where do we go from here?

Let's start with policing. In many cases, mass shooters don't emerge from the ether without warning—they issue threats, they write about their plans on social media, and they attract the attention of concerned civilians and police departments. But again and again, mass shooters have, infuriatingly, fallen through the cracks.

I've written about the many warnings issued about the Parkland shooter, Nikolas Cruz. As the *Guardian* similarly reports:

> Omar Mateen, who killed 49 people and wounded 58 others at the Pulse nightclub in Orlando in June 2016, had previously been the subject of a 10-month FBI terrorism investigation.
>
> Esteban Santiago, who killed five people and wounded eight at the Fort Lauderdale airport in January 2017, had walked into an FBI office in Alaska with a loaded handgun magazine two months earlier and reported having "terroristic thoughts".

Devin Kelley, who killed 26 people at a church outside San Antonio, Texas, [in] November [2017], had been kicked out of the Air Force following a conviction and prison sentence for domestic violence, but the Air Force forgot to notify the National Criminal Information Center so he would be prevented from buying firearms in future.

If future mass shooters—who are virtually flashing red with warning signs—keep falling through the cracks, we have a right to ask why and to insist on rethinking how these warning signs are processed. If human error is so often a cause of these oversights, as it appears to be, then it makes sense to invest in AI and big data approaches to more seamlessly compile, evaluate, and share information across branches of law enforcement.

As John Cohen, a former counterterrorism coordinator for the Department of Homeland Security, told the *Guardian*, the profile traits of mass shooters are well-established: "The objective analysis of what the characteristics of mass shooters are has been done. Building a violence prevention framework has also been done. What we need to do now is work with localities so they can organize themselves, educate the public, and put together the capacity to do behavioral risk assessments. We need law enforcement, mental health and social services to be working together."

I'd add that if we continue to rely on human judgment to do all of the work for us, we're going to wait too long for meaningful solutions. The same violence prevention algorithms that can flag potential perpetrators and victims of gang violence can also combine established profiles of mass shooters with social media data and civilian complaints to flag potential mass shooters. All of that raises the question of what to do with this information. How does a society approach an individual identified as being at high risk of committing an atrocity, but who has yet to commit a crime? I don't have a clean answer. But ignoring that information entirely feels like the wrong one.

NOW LET'S look at the technology side of the problem. Here, I'd like to take a step back to think about a topic entirely unrelated to school shootings: self-driving cars. In the early 2000s, when the idea of a car without a driver was still a concept out of science fiction, the Defense

Advanced Research Projects Agency (DARPA) aimed to bring the sci-fi fantasy a step closer to reality. Self-driving car technology had military applications, but leaders at DARPA could sense its wider civilian applications as well.

Out of that insight came an audacious idea: the DARPA Grand Challenge, a one-million-dollar prize authorized by Congress to anyone who could build a car that would drive itself from Barstow, California, to Primm, Nevada. Over the next year, hundreds of engineers, companies, weekend tinkerers, and university researchers built and tested vehicles. Fifteen finalists were selected, and on March 13, 2004, fifteen autonomous cars set off into the California desert.

Not a single car made it all the way to Primm, but DARPA saw what was clear to everyone who competed: the Grand Challenge spurred a great deal of innovation in a very short time and, importantly, it drew people the world over into the effort. So DARPA decided to double the prize and run the competition again. One year later, a team from Stanford University took home two million dollars when its vehicle completed the assigned course in a little over six hours. The Stanford team was the fastest of five that finished the race.

Google acquired the Stanford team and its technology. A few years later, the innovations and discoveries born in that desert led to the first successful tests of autonomous vehicles on American streets. Today, every car company in the world is working to build its own version of these vehicles.

We can trace much of that progress back to DARPA's Grand Challenge. As Lieutenant Colonel Scott Wadle, DARPA's liaison to the US Marine Corps, puts it, "That first competition created a community of innovators, engineers, students, programmers, off-road racers, backyard mechanics, inventors and dreamers who came together to make history by trying to solve a tough technical problem. The fresh thinking they brought was the spark that has triggered major advances in the development of autonomous robotic ground vehicle technology in the years since."

You already know that I'm very sympathetic to challenges like these. After all, it was a similar challenge from President Johnson—that our nation needed better non-lethal weapons—that ultimately gave rise to the TASER technology. Politicians and policymakers often can't specify the solution to a problem from the top down. But they can create smart incentives, and rely on American (or global) ingenuity to do the rest. It's a decentralized, flexible form of problem solving, and we've reaped rich rewards from it in the past.

The federal government, for instance, didn't build Amazon or Google—but it created ARPANET, which created the packet-switching technology and ecosystem that evolved into the internet as we know it today. The DARPA-funded research enabled the internet, and with it companies like Amazon and Google became possible.

I don't think it trivializes the problem of school shootings to say that, as an especially intractable technological problem, they can inspire hopelessness. And it's in those cases that we can most benefit by turning to the model of decentralized problem solving. Private industry, universities, research institutes, national and military labs, and even weekend researchers may all have an approach to the problem of gun violence in schools that hasn't been considered, an idea that hasn't been tested, a solution that doesn't involve changing laws or even changing minds.

To explore this idea, I spoke to one of the most inventive minds in Silicon Valley. I won't use his name here out of discretion, but he leads one of the most innovative product development labs there is. As we discussed how we might find untried, new approaches to protect schools, we covered many ideas, including more aggressive use of surveillance cameras and AI. Then his eyes lit up.

He picked up a Kleenex box and a pen and began to draw on it. "You could create a small drone that's pre-emplaced in a school in small containers on the ceiling . . . like a smoke detector. When you detect an active shooter, the casing releases, and the small drone is directed toward where the threat was located. You could have a direct connection to a police dispatch center where they could pilot the drone remotely. You could respond immediately, without having to wait for officers to arrive on scene. You could have a small non-lethal electrical weapon, or a tranquilizer dart here and here," he said, drawing circles on the front of the Kleenex box.

He had independently arrived at the same idea for an armed drone that I'd featured in a military application in the Raqqa scenario. Such a system would have several key features that make it worth thinking about in the context of a school. Unlike a gun, it couldn't be compromised by a student who got hold of it. You could encrypt the control codes to make it extremely hard for hackers to gain control, too. And even if they did, a non-lethal drone could cause some problems, but not the mayhem and death caused by a lethal weapon in the wrong hands, or even the right ones. These units could be replaced at relatively low cost, and response times could be near immediate, with no waiting for the police to arrive. And finally, a remotely operated system would not require a teacher (or anyone else) to put their own life at risk to stop the threat.

As we wrapped up, he set down the box and a look of resignation crossed his face. He sighed and said that while these were all interesting ideas, no major Silicon Valley company would touch something this controversial. I knew he was right. I could just imagine the field day the press would have: "Facebook [or Apple, or any tech name] Designing Drones to Electrocute Disruptive Kids in Schools."

Major Silicon Valley companies would likely never touch something as sensitive as employing weapons to protect school kids, even in a country with a terribly persistent problem of disturbed people showing up at schools with assault rifles and killing those kids. His reaction reinforced my belief that only a grand challenge or something like it could provide the necessary cover for tech companies, defense contractors, and independent inventors to engage with this vexing problem.

Would Google, Apple, or Facebook on their own create a skunkworks project to design autonomous systems that could incapacitate a school shooter? Never. But if we had a concerted effort, called for by our nation's leaders and led by DARPA, the Department of Defense, and the Department of Education, might those same titans of tech allow their talent to participate? I suspect they would, and I imagine it would give us a fighting chance against school shootings. Those entities need the right incentives and competition. They need the cover to know it will be safe for their careers to work on the sensitive challenge of how to protect our children from the most vile and violent threats.

We do not know what the precise answer will look like, but we do know that it's worthwhile to ask the question: How can we use technology to stop school shootings? If that question is asked at the highest levels of our country's government, it can bring out the best in innovation. The technology of public safety has evolved rapidly, and with developments in artificial intelligence technology, drones, advanced sensory devices, and non-lethal weaponry, a high-tech solution to the problem of school shootings may be well within our grasp.

I would add that I think the challenge does need to come from the highest levels, because schools aren't, nor should they be expected to be, experts in security and defensive measures. It would be downright strange for my local elementary school to host a competition for ideas on how to protect the school from a shooting. It would be more natural if such a competition were hosted by DARPA in conjunction with the departments of defense and education.

I would urge national policymakers to pause from the usual, intractable debates over guns and take a new tack. Announce a challenge, set clear, measurable goals, and offer rewards valuable enough to enlist our nation's tech innovators into action. Right now, there are few worthier projects for them to be working on.

IT'S IMPOSSIBLE to predict what sorts of solutions would emerge from a challenge like this one. If we knew the answer ahead of time, we wouldn't need to have this conversation. Let me close by pointing out a few promising avenues for research that might yield valuable results if researchers are properly incentivized to explore them.

In Healdton, Oklahoma, public schools have deployed bulletproof storm shelters, allowing children and teachers to take shelter inside their classrooms during both severe weather and active shooter events. What would it take to make such shelters cheap, lightweight, unobtrusive, and a fixture of classrooms throughout the country?

Elsewhere, companies are developing chemical defense systems designed to deter shooters. In Minneapolis, Minnesota, a company called Crotega has developed a threat-suppression system that sprays school intruders with a solution that irritates their eyes, nose, and lungs, in order to disorient and delay them until first responders arrive. Similarly, Nemesis Defense Systems developed a safe yellow spray to disorient school intruders and mark them for police identification. What would it take to deploy systems like these in pilot programs across the country, and what advances in AI and sensor technology would it take to make them deploy quickly, automatically, and with minimal false positives?

Finally, increasing the deployment of non-lethal weapons—stored, as I suggested, in biometric safes—could offer schools a last line of defense. Within the next decade, non-lethal weapons offer will similar benefits to firearms, with considerably fewer drawbacks. There is no foolproof solution to prevent the misuse of non-lethal weapons, any more than there is with guns. The difference is that the consequences of failure, when the non-lethal weapon is well designed, are dramatically reduced. No student should have to go to school in fear that their own teacher will, by accident or in an unthinkable overreaction, shoot them in the classroom. No student, for that matter, should have to live with the fear that their teacher will stun them with a non-lethal weapon, either. But as before, the virtue of non-lethal weapons is that they can neutralize threats almost as effectively as firearms, while transforming life-ending errors of judgment into mistakes that, though painful and damaging, are survivable.

Whether it's remotely piloted non-lethal drones, bulletproof bunkers disguised as playrooms, pepper spray–dispensing CCTV systems, or something else entirely, it's difficult to say which of these solutions, if any, will pan out—or if the solution will look like something that no one has yet envisioned. But announcing a Grand Challenge on School Safety would be a step that could create a breakthrough. It can be a break in the debate, a truce, a path forward that can unite us around a shared enterprise. And when it comes to the remarkable and practical solutions that might emerge, I would never bet against our collective ingenuity.

FOUNDED IN 1994, THE XPRIZE Foundation was formed to spur a competition that could launch a private space travel industry. Its founder, Peter Diamandis, was inspired by the story of the 1919 Orteig Prize, a $25,000 prize for the first nonstop flight between New York City and Paris that spurred Charles Lindbergh to make the famous journey.

In 1996, Diamandis announced a ten-million-dollar prize for the first privately financed team that could fly a three-passenger vehicle sixty-two miles above the Earth twice within two weeks.

The XPRIZE lured competitors into spending well over a hundred million dollars in pursuit of the goal, and led to the founding of the private space industry, which is a multi-billion-dollar industry today. Since then, the XPRIZE Foundation has launched over $140 million in prize purses, driving innovation against grand challenges as diverse as global learning and oil cleanup.

In October 2018, I asked to present the concept for a School Safety XPRIZE at the foundation's annual Visioneering summit. The summit is a fascinating event, where participants present to an audience that votes for the winning ideas.

I confirmed one thing about alternative approaches to school safety: the topic provokes strong responses. Reaction to my proposal ran the gamut from enthusiastic support for fresh thinking on this topic to offense that I would condone the idea of placing any kind of weaponry in schools. One detractor noted, "How can you propose a competition that could lead to wrapping our schools in bulletproof glass, or adding intrusive surveillance or weapons? We need to focus on the underlying problems of mental health issues and guns in America and solve the underlying problems in the first place!" And half of the crowd cheered.

I responded, "I understand your sentiment. I do not condone or accept that it is in any way moral for people to show up at a school and start shooting children. However, I do not believe we can wait for a sea change in the human condition that will suddenly lead to a change of heart for the people who commit these crimes. Hope is not a strategy, and we need a strategy to protect our kids the next time this scenario plays out."

Seeing how the emotions of the crowd impacted the voting, I didn't want to take any chances. So I asked my now eight-year-old daughter to come on stage with me and tell the story of when she came home from school and asked me, "Daddy, will a man come to school with a gun and shoot me?" Her question didn't come out of the blue: she had been through a lockdown drill at her school that day. The training she'd received? If someone came into the school with a gun, she was told, go hide between the backpacks in the coatrack at the back of the room and stay quiet.

As a parent, I was sick to my stomach. Telling my daughter to go hide quietly and hope she doesn't get shot seemed like a dereliction of my duty as a parent. While I couldn't share it with her, my first reaction was, "We would be better off not having these drills at all." We are terrifying tens of millions of kids, with information that has zero utility in potentially saving their lives.

And yet, even with my daughter's teary-eyed appeal, the room split fifty-fifty and my idea lost ... to a proposed XPRIZE for improved battery technology. I mean no offense to batteries and battery makers, but everyone from Elon Musk to Samsung to Panasonic are building megafactories and investing billions to create the next breakthrough in batteries. Because there's a powerful economic incentive to create them, better batteries need no champion or prize. That's a solution the market will find.

The saving grace from this moment: at least half the attendees began talking about school violence in creative, proactive ways. At a minimum, people were willing to stop and ask the question: Is pretending we can change laws or arming every teacher the best we can do to deal with school violence? The CEO of the XPRIZE Foundation has begun exploring the possibilities of such a prize, and I'm hopeful that it comes to pass, so that all parents, everywhere, can send their kids to school knowing they are safe.

CALLING FOR
PROGRESSIVE ACTIVISTS

progressive: \pr*uh*-**gres**-iv\—*adjective*
   1. favoring or advocating progress, change, improvement, or reform,
      as opposed to wishing to maintain things as they are, especially in
      political matters: *a progressive mayor.*
   2. making progress toward better conditions; employing or advocating
      more enlightened or liberal ideas, new or experimental methods,
      etc.: *a progressive community.*

I N THE EARLY 2000s, the Ohio chapter of the ACLU was one of the
most active in the country. Executive Director Scott Greenwood had
spent over three decades as a civil rights attorney who had focused
on police use-of-force cases. In fact, he was the most prolific civil rights
attorney with the most cases against police in the Federal Sixth Circuit. When possible, Greenwood tried to create change within police
departments, but when that wasn't possible, he would often end up
suing police departments for misconduct. He had seen every angle on
this problem.

Following a number of consecutive police shootings involving white
officers and minority suspects, the streets of Cincinnati exploded in

rioting and violence. Three days of carnage and mayhem passed, and by the end, the city instituted curfews. Greenwood filed suit against the police department. "This wasn't the occasional or random in-custody death or officer-involved shooting. This was every day for a few months," he told me.

For him, and the community, the process of policing was badly bro-ken. "Nobody was getting serious discipline and the officer-involved shootings just increased." Greenwood successfully obtained a consent agreement that put him on a civilian oversight board with a role in set-ting policy for the Cincinnati Police Department.

He started by going to the city's legal department and then to the police chief, saying, "We have to stop this. There has to be a better way. We know that there are ways to use less force and my promise then was that I will not file another individual lawsuit against the police depart-ment as long as you are constructively engaged with me and with the ACLU of Ohio in addressing these problems." The conversation began a collaborative process between Greenwood and the city.

When Axon first approached the Cincinnati Police Department, we were told that we'd have to present to the oversight board and receive their approval for the TASER devices. Based on prior experience, we knew this would be difficult. At our first engagement, Greenwood was clearly skeptical. For him, giving the police a new weapon seemed like a terrible idea. "A police force that used too much force on too many people at the time . . . we didn't think should be trusted to use yet one more weapon," he said.

So he asked a number of pointed questions, and he expressed skep-ticism that more or newer weapons would somehow lead to fewer deaths. He was very direct: "You want to sell these to this agency. I'm going to block it unless I know how they work. I want to know as much about these weapons and how they're used as any person other than the engineers who designed them."

To his credit, he agreed to come to a training event and see how the weapon worked. For some context, Greenwood is a soft-spoken intel-lectual with a lawyer's eye for words and logic. So it surprised us a bit when he agreed to volunteer for a TASER device exposure. We shared both technical and field data about how TASER devices worked, and case studies where they significantly reduced injuries to both police and the public. He took all of that information and decided that the practical experience of undergoing a TASER device exposure would help him understand the effects of the weapon and what impact it might have in the community.

In the end, after studying the research and having a TASER device used on himself, Greenwood approved a pilot program. When that pilot proved successful at reducing police confrontations, he approved widespread deployment of TASER devices to front-line officers. He also led an effort to track their results and measure their impact. Just because he had approved a pilot didn't mean he wasn't going to remain a watchdog.

As a result of these efforts, Cincinnati went from being torn apart by police shootings and subsequent riots to a period of relative tranquil-ity. Greenwood himself described the impact in the following terms: "Cincinnati PD didn't fire a shot for twenty-seven months. They went from eighteen dead young African-American men over a period of not too many years to zero shots fired. Therefore, zero people were killed. That is an absolute, out-of-the-park grand slam." Looking back on the work he did, Greenwood commented, "It was one of the most impactful things I have seen to improve the lives of people living in communities beset with violence and aggressive policing."

At the national level, the ACLU has been a major force behind the nationwide adoption of body cameras by police officers. Body cameras, like any technology, are imperfect. They are not a panacea. But there's reason to believe, as I discussed earlier (see chapter 9), that officers who know their actions are being recorded will act in a more account-able manner. And body cameras allow the public to keep tabs on those we entrust to commit violence when necessary in our name, which ought to be a must in any democracy.

Real progress doesn't occur all at once. It happens piecemeal, and it often requires compromises and setbacks. Working in the field of tech-nology, that's something I know from painful, first-hand experience. But it's a lesson that activists interested in making real progress should keep in mind as well. Those who stand to make the biggest difference in our world must know that making a difference is usually measured in fits and starts—but that progress generally depends upon people being willing to say what they are for, not only what they are against.

The lesson from Scott Greenwood's experience isn't that TASER weapons are good. It's that activists can play a constructive role, that they can help both technology providers and police be more thoughtful and careful while also fulfilling the mission that their organizations believe in.

Activists who collaborate rather than critique can take some heat. And Greenwood's experience has been no different: he has dealt with criticism for his willingness to work with both police leaders and technology companies. When he began working on constructive solutions, it upset some who saw his collaboration as selling out. He especially upset the diehards who see collaboration as weakness or complicity with "the system." But few of those critics can point to anything like the positive impact that Greenwood has achieved. He can offer numerical, incontrovertible proof that he's changed his community for the better.

These can be highly polarizing issues. Those who strongly favor law enforcement can take the side of police and the military uncritically, even in cases where they're in the wrong. And in the same way, activists can paint police or the military with broad brushes—believing that all of them have a predilection for violence or racism—even in situations where officers and soldiers show restraint and compassion.

Social progress relies upon solution-oriented leaders willing to reach across those ideological divides—to put their egos to the side and find the shortest route to results. No one has a monopoly on the right answer, and as the CEO of one of the more influential companies in this space, I've seen enough to know that the best answers tend to come from many different directions and perspectives.

I DON'T WANT to understate what activists and organizers are up against when they do their work. Making change means taking on entrenched systems of power—economic power, political power, or cultural power. It's hard work, and it probably means losing more battles than you win. Activists might not like to hear it, but I identify with those struggles, because making the TASER dream a reality has meant going up against entrenched systems, too. And I realize that today, some activists may view me and my company—ironically—as representative of the entrenched system.

But activists are up against something else, too—human nature. Humans have a tendency to define issues in adversarial terms. For example, if you are a peace activist, it's easy to become opposed to the military. If you are a civil rights activist, you can be implacably against any police. If you are an environmental activist, you might be resolutely against the energy companies.

Those are easy positions to take; the talking points have been written and rehearsed for decades. What's much harder is to stay focused on the sometimes-abstract goal you're after. It's harder, in other words, to be "pro peace" than it is to be "anti-military." It's harder to be "pro civil rights for all" than it is to be "anti-police." That's why I believe that most protests end up directing their energy against something or someone rather than promoting an idea or concept. Ideas and concepts don't speak to the gut in the same way that opposition does; emotion brings people to the barricades.

But those same emotions affect not just how we speak to other people about the causes we care about, but also how we think about them. They lead to blind spots in our thinking. In chapter 2, I discussed some of the cognitive biases that keep us from embracing change, even when it's change that will, on the whole, make us better off. One of the most important of those biases is the tendency to make the wrong comparisons when we're evaluating potential changes. Rather than asking, "Is this an improvement over the status quo?" we all too often ask, in effect, "What are the flaws?"

As a result, once we identify any flaws in a proposal—and every proposal has flaws—we're primed to reject it. That's the case even if it's less flawed than its real-world, status quo competitor. Most of us have a built-in conservative bias against change, and we have to fight to identify it and correct for it if we actually want the world to change for the better. Unfortunately, that's as true of activists, even the most progressive activists, as it is of the rest of us.

Think, for example, of the phenomenon of NIMBYism ("Not in my backyard"). In cities like San Francisco, activists have long stood against the kind of new construction that could bring down housing costs. They've identified plenty of ways in which building new housing falls short of perfection: it disrupts existing neighborhoods; it enriches developers; and, of course, no one likes construction and extra traffic. But their efforts also exacerbate the problem of San Francisco being the most expensive city in the United States, one that's inaccessible to the working class and, increasingly, to the middle class. It means less social mobility, less opportunity, and less change—all of which are pretty unprogressive outcomes.

IN AN INCREASINGLY polarized world, the assumption is that progress has to be all or nothing. Trade-offs, compromises, and sacrifices in the pursuit of progress aren't allowed, and any accident or misstep is treated as a damning condemnation. And that's troubling, because trade-offs, compromises, and sacrifices are most often the way that progress is achieved.

I don't begrudge people who have a big goal connected to a deeply moral purpose. If anything, I worry about people whose big goals are disconnected from a moral compass. But to engage the analogy a bit, on a long journey, a compass alone isn't enough. As it was put wisely in *Lincoln*—a film whose central question is what kinds of compromises President Lincoln was forced to make on the path to achieving his goals—a moral compass will "point you true north from where you're standing, but it's got no advice about the swamps and deserts and chasms that you'll encounter along the way. If in pursuit of your destination, you plunge ahead, heedless of obstacles, and achieve nothing more than to sink in a swamp—what's the use of knowing true north?"

Activists often have a powerful, admirable sense of true north. It's what motivates their work, and it helps generate the moral conflict that attracts attention to their cause. But a moral compass tells us very little about how to act strategically to achieve moral goals. So environmental activists can become laser-focused on the sins of the energy companies, but they are liable to miss the broader strategic picture of how they may positively influence those same companies to change their behavior. They may miss opportunities to promote technologies that might help to dramatically reduce pollution. Instead, they focus adversarial energy on companies that pollute, even when there may be other, viable ways to work with the companies to help both the activists and the companies achieve their ends.

I recently spoke to the CEO of a company that sold software to energy companies that helped them reduce pollution by improving their operations. Because of his work, he was often invited to speak at environmental conferences or would find himself on recruiting trips to colleges. He would open his presentations by explaining how his company both helped the environment and made energy companies run better.

But the idea of both sides winning fell flat with his audiences. Many of the attendees would tell him that they viewed energy companies as evil enterprises run on a pure profit motive—that any action taken to help them was, by definition, an action that harmed the environment. He could prove this wasn't true based on the work he was doing, but his argument fell flat. So he learned to edit his presentations to focus only on the pollution he was reducing—and not mention that he was helping the companies' bottom line at the same time.

In the same manner, some peace activists' disdain for the military or the police can prevent them from engaging constructively with those institutions in order to change them for the better. The fate of the Active Denial System—the non-lethal heat weapon that I discussed in chapter 7—is an excellent example of the way this pattern has stood in the way of real, progressive change. The ADS, to recap, was a non-lethal weapon that used directed energy to disperse crowds. When it was first developed, I believed—as I still believe—that it had the potential to begin a shift toward a military that could achieve its missions with significantly less loss of life. But so far, that potential has gone unrecognized.

I believe anti-war activists, who were a major force lobbying against the adoption of the ADS, made a serious mistake. They won the battle against that specific technology, but they set back progress toward a world that is less deadly—a world that they surely are fighting for. Activists focused on the frightening aspects of a new technology (*The military is building a death ray!*) at the expense of the more hopeful and realistic aspects (*The military is finally moving away from lethal weapons!*). It was easy for them to focus on what might go wrong (*The ADS might malfunction and kill by accident!*) at the expense of a fair comparison with the status quo (*The ADS could replace weapons like the M16 that kill or maim with every use!*).

In that situation, the activists chose a stance of moral purism over pragmatic progress. If you're categorically opposed to the military, anything the military does is wrong. Victory becomes anything that slows the military down. That stance prevents you from driving the military to do better—in this case, by investing in weapons that don't kill, and by progressing toward a future where there is less collateral damage and fewer innocent lives lost.

In defeating ADS, activists achieved a pyrrhic victory: they killed more than just ADS; they have nearly killed the concept of non-lethal military options. The military has seen no return on its investment in non-lethal technologies, besides public relations problems. The institutions promoting non-lethal approaches are getting defunded and deprioritized to the point of irrelevancy. That is the worst outcome possible for everyone involved—the activist community, the military, and humanity at large.

AN UNINTENDED consequence of protest is that the institutions that activists protest against can become less responsive to their pleas. If leaders of organizations believe that they will be criticized no matter what they do, then why not continue the status quo rather than try something new?

International activists have gone after non-lethal weapons for, perversely, making warfare more deadly. "We had a real complaint from some on the humanitarian organizations that this is going to make warfare too easy," General Anthony Zinni told me. "That by making war less lethal, you were making it more attractive and thus more likely to happen. So therefore we should not pursue non-lethals, and we should keep lethal weapons. I mean, the logic just defies rational thinking." Even in reasonable discussions about the future of acoustic weaponry or laser-based weaponry, organizations would launch pre-emptive strikes. "You're going to blind people, they'd say," Zinni remembers. "Even though the technology was going after temporary incapacitation, but they didn't see that."

I've seen this first-hand in some of my interviews with military experts. They talked about how the PR risk of deploying the ADS was a significant deterrent to its use, and further investment in safer and more humane options has dried up. I've seen this same pattern play out with TASER technology. Anti-police activists have shouted from the rooftops that, because electricity has at one time or another been used to torture human beings, any weapons that use electric charges are by definition dangerous and shouldn't be put in the hands of cops.

We've pointed out that life-saving technologies like pacemakers and defibrillators use electricity, too, as does any home appliance or even the laptop these words are being written on. But the image of electricity-as-torture-method is a powerful, visceral one. Perhaps inadvertently, the activists have ended up promoting the non-electric, lethal weapon in the hands of every cop in the United States: the gun.

During the height of the controversy over TASER weapons, a police chief told us bluntly that the department was better off from a public relations perspective using their guns. While vastly more dangerous, for both civilians and police, guns were well understood and accepted. "If we shoot someone with a gun, that's something the public understands. But if we shoot them with a TASER device and something bad happens, we will be villainized," he said. As I keep pointing out, we don't judge the status quo and disruptive innovations on a level playing field. We hold the latter to a far higher standard. Stun guns, the police chief told me, would spark a public backlash in a way that sticking with the status quo, bullets, would not.

It's been frustrating to see activists who share our goals come out against us. Rather than being focused on our common goal of ending police killing, many activists demonized us for developing alternative weapons designed to reduce human injury and deaths. Several leading human rights organizations led an intensive, multi-year campaign to ban the use of the TASER technology on the basis that it might be used by thuggish dictators to torture—as if restricting the tools would prevent the practice.

Calls for TASER technology bans got newspaper headlines— which is one of the most visible impact metrics for non-governmental organizations (NGOs). If you measure success by media impressions, the campaign against TASER weapons was a success. NGO officials were quoted in numerous front-page stories around the world. They changed the public mood about TASER weapons, and their campaign derailed programs to deploy TASER devices in a number of countries.

But should that constitute a success for human rights and human safety? The upshot of their effort is that more police around the world carry weapons that are lethal, with no meaningful alternative. By demonizing TASER technology, these well-intentioned groups were, inadvertently, lionizing guns. Focusing on the ways that non-lethal weaponry can go wrong leaves us a world of highly lethal weapons as the status quo.

The result harms us all: by my estimation, activists successfully delayed the widespread deployment of non-lethal weapons in many countries by at least a decade. When I spoke to one activist researcher about the matter, her response was revealing. I had offered the idea that perhaps they could work with my company and police agencies to help institute best practices. Unlike electricity from wall outlets or car batteries, every TASER device has a built-in audit log that records every use, and TASER devices could be outfitted with cameras that could record every use to ensure accountability.

Would they want to join forces to help create oversight guidelines and mechanisms to help ensure these technologies would be used in a manner to augment the benefits and reduce the risk of misuse? Would they want to see some of our data and understand our work? It would send a powerful signal if a technology company, police accountability activists, and police leaders worked together to drive meaningful change with new oversight mechanisms. The researcher thought for a moment, then responded, "If we become a part of designing the system, then we could lose our credibility as an independent watchdog." This frustrates me to this day. What she was saying, in effect, was that remaining a critic was more important than fixing the problem she was criticizing.

I WANT to be clear about one thing: human and civil rights organizations have done remarkable work on behalf of political prisoners and advancing human rights around the world. Their work in those domains is admirable. I believe that these activists weren't malicious in launching campaigns against us; they were just misguided. Their decisions can be chalked up to the same bias that keeps us from exchanging an imperfect status quo for a less-imperfect change.

NGOs face the same pressures that all large organizations, for-profit or non-profit, do. Just as corporations fund their operations by generating sales, non-profits raise donations by generating public interest and support through calling attention to outrages. A viral outrage reliably generates more interest than a position of "This is bad, but on balance it may lead to an improvement over a status quo that is also bad. Let's keep an eye on it and see." And as a result, NGOs can become conservative forces in spite of progressive intentions. The urge to criticize the actions of police agencies or companies or militaries can undo positive actions, slowing down all change, positive and negative.

The activist community in San Francisco has followed a similar, stultifying playbook. Following several fatal police shootings in the early 2000s, the San Francisco Police Department sought approval from their civilian police commission to deploy TASER weapons as an alternative. Activists staged raucous protests at every hearing, on at least one occasion chanting "Fuck you, Steve" to drown out testimony from one of the experts before the police commission.

Those tactics worked: as of 2018, the SFPD has yet to deploy a single TASER device. That makes it an anomaly. As of this writing, San Francisco is the only major city in the country that does not give its officers access to TASER weapons. The reason is clear: a highly active anti-police activist base has prevented progress by resisting all efforts to introduce alternative force options.

The results are plain for all to see: since 2000, there have been more than a hundred officer-involved shootings in San Francisco. Many of those have resulted in the suspect being shot and killed. In most of those interactions, I—and others—suspect that a non-lethal option could have prevented the situation from escalating to deadly force. And yet, police officers in San Francisco have no choice but to use and fire their guns in these situations.

IN THE COURSE of developing the TASER technology and the field of non-lethal weaponry more generally, I've gotten to know a broad cross-section of the public. I've worked closely with scientists and engineers, with politicians and public officials, with law enforcement officers and military personnel, and with activists in the non-profit sector.

One thing I try to keep in mind is that essentially no one, in any of these groups, gets out of bed in the morning and sets out to make the world a worse place. They all want to make a contribution. They want to make progress. It's essential to remember that, even when we have serious conflicts over the right thing to do, on a deeper level we're on the same side. We all set out to make this a safer, more peaceful world; we just have different perspectives about how to get there.

I particularly try to remind myself of this when I come into conflict with activists who have a different vision of the future than I do. Like them, I want this to be a less violent world. Like them, I want to be a "progressive"—in the original sense of that term, meaning someone who inspires and drives change. Even when we disagree—and we do disagree—I firmly believe that we see ourselves in the same way.

It's easy to decide what you are against. It can be much harder to define what you are for and then to fight for it, even in the face of critics. Protests can create the energy to drive reform. Constructive collaboration converts this energy into results. The world needs activists who want to change things for the better. Your passion can catalyze reform and lead to constructive solutions that propel true forward progress.

13

OUR BIGGEST
PROBLEMS REQUIRE OUR
BRIGHTEST MINDS

I F YOU'VE HEARD about the wildfires that sweep the American
West, you've probably learned that putting out a wildfire isn't like
putting out a house fire. A house fire is challenging, of course, but
as long as the fire is contained and the nearest hydrant works, there
are plenty of resources to solve the problem. The hydrant will give you
enough water to spray at the house until the fire is out.

Wildfires are a different beast. The worst can consume hundreds of
thousands of acres. Even if there's theoretically enough water available
to douse all of those acres, trying to get it there is counterproductive
at best and impossible at worst. When the fire is that big, firefighters
have to pick their spots. They look at which sections of forest are driest
and most likely to catch fire; they look at forecasts of prevailing winds
to get an idea of where the fire might spread next. And then, rather
than trying to put out the whole fire at once, they direct their efforts
to the most critical areas. They clear vulnerable areas before the fire
can reach them, or even set smaller, contained fires to control the path
of the big blaze. If they direct their resources wisely, they can bring

the wildfire under control. But they have to pick, and some choices are more optimal than others.

I like to think of a society's supply of intellectual talent as the equivalent of those firefighting resources. Wherever those resources get directed—through a combination of market pressures, political influence, and cultural signals—problems get solved; fires get put out. If a society is wise about where it directs its limited resources of talent, it can put out more and bigger fires. If it makes bad decisions about talent allocation, it can put out a bunch of insignificant and out-of-the-way fires, while the blazes of the biggest problems burn out of control.

Think of where a society prioritizes sending its talent and smarts, and you'll know a lot about the problems it's good at solving. In ancient Rome, any ambitious young man who wanted influence, fame, and a political career had to spend time leading troops in the army. Unsurprisingly, the Roman army was second to none. In medieval Europe, the clearest route for ambitious people who wanted to rise above their social station was through the Catholic Church. Again, unsurprisingly, the church was home to the most skillful diplomats, the most learned scholars, and the most accomplished artists. In China, for centuries up to the modern era, the road out of the provinces for ambitious youngsters meant mastering the Confucian classics, acing government exams, and entering the service of the emperor. And again, the imperial court earned a near-monopoly on China's best and brightest. You can tell a lot about what a society values from observing where it sends its talent.

As far as I'm concerned, today's equivalent of the Roman army, the medieval church, or the Chinese imperial court is Silicon Valley. Its wealth, opportunities, and entrepreneurial culture make it a magnet for talent from all over the country and the world. In itself, that's not a problem. Something like Silicon Valley will probably exist in every culture; talented people everywhere are drawn to the sources of wealth and power.

What concerns me, though, is that having concentrated so much of our intellectual resources, the tech industry is systematically disengaging from the problems that concern us the most. Sure, you'll find change-the-world patter in pretty much every start-up pitch you care to listen to. And when the likes of Google, Facebook, and Amazon put their minds to solving a problem—whether it's self-driving cars, connecting billions of people to each other, or organizing the world's information—they can achieve remarkable things. But from what I can see, much of the tech industry is turning its back on some of our most significant social problems.

I think that's especially the case when it comes to the social problem of violence. Of course, that's one I'm going to be attuned to, because it's what I spend most of my time thinking about. But it's not as if bringing down the United States' rates of violence is some sort of niche concern. I continue to find it tantalizing to think of what might be achieved if Silicon Valley treated the problem of violence with the vigor it brings to new gaming apps, social networks, or business software.

Turning their backs on the problem of violence might make the people of the tech industry feel better about themselves. You don't have to answer hard questions about how to approach seemingly intractable problems if you simply walk away. But that doesn't make the problems disappear. If you aren't part of the solution, you are part of the problem. When it comes to violence, Silicon Valley should be part of the solution.

IT'S NOT JUST that Google and Facebook have decided to ignore the problem of violence. It's that their preferred methods of addressing the problem are woefully insufficient.

In the face of growing public concern over gun violence, the major tech platforms' one action has been to ban all ads for all self-defense systems, including guns and non-lethal weapons. At first blush, that might sound reasonable—less gun advertising would mean fewer gun purchases and therefore less gun violence. But it's not so simple. Did the tech platforms ban these ads after rigorous research into the connection between gun advertising and gun violence or simply as a means of attracting some good PR? I strongly suspect the latter. Will weapons sales really drop if Google stops allowing weapons ads? Will gun buyers really put their guns away in response to a Facebook policy? I'm doubtful, but more to the point, I don't believe that Google and Facebook themselves know the answer.

Furthermore, advertising can be a powerful force for social change. By cutting off all weapons advertising, the tech platforms are cutting off the chance for this force to reshape a hidebound industry. Advertising doesn't just tell us about new products on the markets; it shapes what we want, what we consider desirable, and what we consider normal. Think of Apple's revolutionary ads for the Macintosh and then for the iPod, which convinced millions of us that personal computers and portable music players were not just for the technologically sophisticated, but could improve life for everyone. Or think of how advertising across industries has increasingly foregrounded multi-racial or same-sex families, reminding us that there are many ways to be a "normal," loving family.

Imagine an alternative case. Imagine if Facebook and Google sought to address climate change by banning ads for cars—all cars. This would obviously be ineffective. Neither the rate of cars purchased nor of cars produced would change. Outlawing those ads could also be counterproductive. It would prevent companies from promoting new vehicles, such as hybrids or all-electric vehicles, and a powerful force to shift consumer tastes in a direction that could help address climate change would be lost.

Advertising can and does change our world, and that includes advertising for weapons and personal safety. To achieve critical mass, non-lethal weapons like TASER devices or advanced technology like "smart guns" that only fire for a designated user need an educated public aware of the existence and benefits of these products. That's the only way they can challenge old incumbents and meet public demand for safer weapons. I don't think that an ad ban from Google, Facebook, and others will have any real effect on gun sales, because, simply put, the public already knows what a gun is and how it works. But I do believe that many people who might have researched and bought a safer alternative are currently unaware that these alternatives exist and continue buying guns as a result.

The number one question we get from private citizens is: Are TASER weapons legal to own? (The answer: Yes, in most states.) Advertising exists to answer that kind of question. And we have been categorically banned from the most extensive advertising networks. The ad ban was obviously well intentioned. But its unintended consequences include holding back the spread of new ideas, such as non-lethal weapons—one of the most promising routes for reducing lethal violence.

Here's another, non-advertising example. Google's employees are agitating strongly to prevent any Google technology from being "militarized." When Google bought Boston Dynamics, a leading robotics company, it quietly refocused the company on building consumer-oriented robots and away from the Department of Defense. While killer terminators certainly raise concerns, there are numerous uses for military robots that could remove human operators from dangerous non-combat roles, such as using robots to deliver supplies and equipment, even for humanitarian supply missions. Many of the lives lost to improvised explosive devices (IEDs) in the past decade have been in convoys delivering logistics support. Would it be morally offensive to take these convoys out of harm's way? As a father whose son survived a road-side bomb attack on his vehicle while serving in Afghanistan, I bet you can guess my opinion.

Google has also moved to prevent the military from using its AI technology. Again, these might seem like laudable, peace-loving, don't-be-evil moves. But artificial intelligence could reduce innocent deaths and prevent lethal violence. It's far from clear that keeping better analytics tools out of military hands will save lives. And while staying away from hard problems makes your media relations strategy easier, I am far from convinced it makes the world a better place.

I CONTEND that many of these policies are comfortable, not laudable. Google and Facebook's refusal to engage with the problems of violence doesn't mean that those problems will go away; it just means that other people will be trying to solve them. And if we assume these companies are highly effective at hiring the best and brightest, this leaves some of our best talent walled off from solving our biggest problems.

I've argued in this book that the US Military, which is famously conservative when it comes to fundamentally rethinking its doctrine and weapons systems, has neglected and underinvested in the potential of non-lethal technology. Each year, the military gets better at killing people, but each year, it lets slip opportunities to solve problems without killing.

This is where Silicon Valley could play a transformative role. Google is famously not a stuck-in-its-ways institution; it is synonymous with innovation. Imagine a cadre of Google engineers working on solutions to make the military more effective and less deadly. What would result from that kind of collaboration? What could we learn about warfare that we do not know today when some of the world's brightest minds work on the issue?

Importantly, if the smartest people choose not to work on police, military, and defense technology, that necessarily means that those problems will be confined to a less-smart crowd. That can't be a good thing. And the withdrawal of our best and brightest from the field of military technology doesn't happen in a bubble. You can guarantee that the smartest people in Russia and China aren't turning away from work on military technologies. That's exactly where those societies are sending some of their best and brightest. Will the world be better off if Russian and Chinese military technologies pull ahead of those of Western democracies?

We may well be seeing the signs of this talent investment in the offensive cyber-capacities being deployed today by the Russians in election meddling and the Chinese in industrial espionage. If, indeed, the Russian interference in the 2016 US election swung enough voters to change the election result, it could arguably have been one of the most effective military campaigns in history, and one executed cheaply and without killing or physical damage. Similarly, information is coming to light that Russian operatives have been successfully fanning the flames on both sides of the political spectrum by promoting inflammatory rhetoric. It is especially disappointing as I watch family gatherings devolve into vitriolic political arguments, realizing that the anger stoked between us is precisely the intended effect from foreign adversaries. What better way to assail the most powerful nation on Earth than to set in motion a cycle of internal strife that threatens to destroy it from within.

"America, we have a problem," said Representative Jackie Speier, a California Democrat who sits on the House Intelligence Committee. "We basically have the brightest minds of our tech community here and Russia was able to weaponize your platforms to divide us, to dupe us and to discredit democracy." Like it or not, the major tech companies are already embroiled in the struggles between governments and the rivalry between democracies and autocracies.

It would be trite to point out that the Americans, French, and British looked away from war-making in the 1930s, and the result was catastrophic. However, history has shown that ignoring dangerous problems doesn't make them go away; it gives them space to fester. In this respect, I think Amazon CEO Jeff Bezos has got it right. As he put it in 2018, in response to employee protests at Google and other tech companies, "If big tech companies are going to turn their back on [the] US Department of Defense, this country is going to be in trouble."

And, he might have added, other countries will benefit. China is gaining rapidly on Silicon Valley, including in areas like facial recognition technology: solutions like Megvii's Face++ are in the lead over American counterparts. Meanwhile, employees of Silicon Valley firms are demanding that their employers refuse to sell facial recognition technology to law enforcement.

What happens when Chinese firms outpace US firms on technology? Law enforcement will buy from the Chinese. And while an American firm would have to grapple with the ethical and technical challenges of bias in algorithms, inaccuracies in data gathering, and privacy in data storage, the Chinese counterpart may not be as attuned to those issues. In other words, American companies that choose not to participate in a given field make it possible for other companies from other parts of the world to dive in, and to do so in ways that might not be consistent with the democratic values our society cherishes.

We need our best minds on these issues. While there might be legitimate concerns about overinvesting on technologies in public safety, I think we're at the other end of the spectrum right now: the intellectual firepower in Silicon Valley is only too happy to leave the messy business of warfare, police work, and public safety to other people. Of course, we don't want our smartest people running rampant and building the Terminator or Skynet. But aren't our smartest people the very ones we want working to ensure that military technology remains firmly under human control? Don't we want those people to work as hard as they can to help keep our society safe, to reduce violent crime, to put an end to the drug war? And as the focus shifts from the old world of killing to the new paradigm of winning conflicts with minimal loss of life, we need our best at the forefront.

While Silicon Valley is using virtual reality and wearable tech for entertainment or to help you take unicorn selfies in Snapchat, Axon is using VR to create immersive training for police officers to learn how to de-escalate situations involving people in mental health crises. Building tools that teach empathy and de-escalation to law enforcement does more to protect lives and civil liberties than working on entertainment apps.

Rather than build AI technology that might be able to stop a terrorist attack before it takes place, big tech firms are busy applying artificial intelligence to serve up clever advertising. This was captured in a quip by Jeff Hammerbacher, founder of Cloudera: "The best minds of my generation are thinking about how to make people click ads." While Google and Facebook continue trying to convince brilliant software engineers to work on catchier YouTube ads or cuter Instagram filters, a new generation of companies will rise in importance by showing the courage to solve the more pressing societal problems of our age.

I am confident in our society's ability to solve the problems on which it chooses to concentrate its resources—to put out the fires we want to put out. But I am deeply concerned that we are not aiming at the right fires. Silicon Valley has a major, transformative role to play in making our world less deadly. And frankly, it should disappoint all of us to see the Valley shirk that opportunity in order to develop better advertising engines or more addictive social apps.

That must change. Reducing and eliminating deaths in conflict is an important measure of social progress—on par with the most important challenges facing mankind today—and it deserves the attention of our best minds.

14

GOVERNMENT IN A
WORLD WITHOUT PRIVACY

WHAT IF PRIVACY were no longer possible? What if governments were omniscient, with access to all of the details of our lives? How would we rewrite our laws to accommodate such a world?

Of course, that situation is far from a utopia. Privacy is worth protecting, and I don't think we should give up the fight for it. But we also need to face facts: our hyper-connected world makes privacy less and less viable—and less and less valuable—every day. Rather than thinking reactively about threats to our privacy, we need to get ahead of those threats. We need to ask how our lives, our laws, and our governments should change in a world where privacy seems to be shrinking by the day.

We want governments that are both data-driven and well informed, committed to protecting our lives and liberties. We cannot expect, nor do we want, our government to operate in ignorance or in a state of technical incompetence. We cannot protect the society of the 2020s by using the technology of the 1970s. As facial recognition and ubiquitous data suffuse society, changing everything from how we log in to our smartphones to how we connect with like-minded people online, it is unreasonable to expect we will retain the same experience of "privacy" that we experienced in a world before the internet. Returning to the past is not an option. But thinking proactively about the future is.

We can try to regulate away the government's ability to use data to achieve the goals we set for it as citizens, but all too often, those efforts are likely to prove futile or counterproductive. We are living in a data-driven world. Modern corporations are expected to use data to better serve their customers. We should expect the same from our governments—to use data to keep us safe and to ensure fairness and equity. It's not realistic to push governments to be oblivious to data. But it is realistic to ask them to use data for the public good. In this chapter, I'll explore some of the ways in which governments are drawing on new tools to protect the public while minimizing unwarranted intrusions of privacy.

OVERSIGHT AND DATA SEARCH CONTROLS

When government agencies make use of data to carry out their functions, citizens should insist on strong, judicious oversight. Agencies should log search queries into their databases, and they should keep strict controls on access, to ensure that the people viewing and analyzing sensitive information do so with proper approvals and sufficient justification. I stress *justification*, because the same search of the same data can have wildly different meanings depending on the purpose for which it's carried out. Consider a search through a license plate database: it's one thing when it's used to locate a kidnapper and something else entirely when it's used to identify and track vehicles at a peaceful protest. I wouldn't want to give up the first use out of fear of the second, but I also wouldn't want to ignore the risks of misuse.

A balanced approach would put appropriate controls in place to enable the use of data for public safety while minimizing the risk of its use against the public interest. Fortunately, we live in an era in which software can be designed to do exactly that: we can, for example, create inalterable audit logs that show exactly when data was accessed and by whom. Many such systems today require the user to enter a reason for the search, as a way of ensuring additional oversight. For sensitive data or more exhaustive searches, a pre-search approval process can be implemented, involving different levels of authority depending on the sensitivity of the search. Just knowing that these systems are in place can deter bad actors within government agencies from abusing their power.

I've learned about these systems from personal experience. My company, Axon, hosts a large data set that now includes almost fifty million gigabytes of often-sensitive police body camera videos. Our engineers have worked hard to ensure that the data is stored so that only authorized personnel from each agency can access their agency's digital evidence. And every time the data is accessed, that instance is recorded in an inalterable audit log. Even though I am the CEO of the company, for example, I have no way of accessing video data from agencies that store information with us.

Access controls and search logs are important factors that government agencies should require—both of themselves, and of private sector partners who provide data hosting services. But more generally, it's not only privacy-compromising technology that's evolving; so are the controls and oversight that protect that technology from abuse. We have the ability to secure data in ways we could never have imagined. That ability can help us think more creatively and less fearfully about the use of data in the public interest.

THE ENCRYPTION-PRIVACY DEBATE

Following a 2015 mass shooting in California, the FBI made headlines when it attempted to force Apple to unlock contents of the iPhone owned by the shooter, Syed Rizwan Farook. The iPhone case presented an interesting challenge: on the one side, there was a legitimate request from the FBI to help extract information from a smartphone that belonged to a dangerous person. That phone likely contained information that would help investigators further their case and understand the killer's motives. On the other side, a leading technology company argued that creating a tool to extract that information from the phone would require placing every iPhone on the planet at increased risk of a cyber-attack and privacy breach.

As with so many polarizing issues, people picked sides, leaving no room for compromise. Either you were for privacy, which meant total encryption and no cooperation with government authorities, no matter the public safety threat. Or you were for unrestricted government surveillance.

Enter into this debate Ray Ozzie, former chief software architect at Microsoft and an early software innovator who helped create Lotus Notes, one of the first blockbuster software products. As reported in an April 2018 interview in *Wired* magazine, Ozzie was troubled that the debate had become increasingly politicized, and that many in the tech industry considered it an unsolvable problem. He sought to address the impasse by creating a solution that could solve the problem of protecting users' security while ensuring that government could access data under exceptional circumstances.

Ozzie turned to a two-party solution: technology companies would use the same methodology they used to "sign" critical software updates, such as updates to the iOS operating system on the iPhone. These software updates authenticate the software that upgrades your phone.

If a hacker were to gain access to the keys to the update system, they could install a software update on your phone that would give the hacker complete control. Hence, the handling of the keys is critical to maintain the security of the operating system, and every smartphone in the world. It's possible to use the same method to authenticate access to encrypted data, without introducing any new security risks. Technology companies would simply have to agree to cooperate with public safety agencies in cases in which a court deems cooperation necessary.

There remains some debate over whether such a system could operate securely on a large scale. And, of course, the privacy questions grow even more pressing when the governments that ask to cooperate with Apple are authoritarian ones. But Ozzie's approach still suggests the possibility of transcending old and stale privacy/security debates. It suggests that, even in an age of polarization, it's possible to strike a dynamic balance between those two goods. That's progress in the best sense.

SURVEILLANCE AND SENSORS EVERYWHERE

We live in a camera-saturated world. Cameras are on our phones, our ATMs, our doorbells, our streets. That's just scratching the surface of the many mechanisms by which our movements are being watched and taped.

Today, the handling of these video sources is haphazard at best. But imagine if, knowing that cameras were as much a part of our lives as stoplights, we figured out a way to consent to the use of video data by law enforcement in extreme circumstances. One example is the app Noonlight: smartphone users set up the service to hold their personal data and transmit it to local 911 dispatchers in an emergency. We could imagine similar services to connect private cameras to police forces. As discussed earlier (see chapter 5), Chicago Police Department already has tens of thousands of private cameras feeding into its surveillance center. Ring (now owned by Amazon) and other home surveillance providers are offering services to share video within neighborhoods or directly to law enforcement.

Surveillance data from these sources can be used in a way that keeps our streets safer, while also protecting privacy. Perhaps the best example of how it might work comes from audio, not video, surveillance. ShotSpotter is a network of sensors installed in many major cities that helps detect the sound of gunshots. If a gunshot is picked up on a network of microphones placed around a city, the location is beamed to police agencies, who can respond rapidly. This technology is not to be underestimated: many gunshots go unreported in high-crime neighborhoods, and ShotSpotter is often the only method by which police know that a shooting has even taken place.

But as helpful as the technology can be to solving and preventing crimes, it comes with a downside: it turns out that the microphones required are also sensitive enough to record nearby conversations. That raises the possibility of surreptitious recordings that might violate wiretapping laws. To mitigate such concerns, ShotSpotter microphones automatically stop recording four seconds after they are triggered—enough to pick up gunshots, but not enough to record a conversation.

A major challenge for video surveillance that protects the public but is still consistent with privacy will be developing controls that work in

a similar way in more difficult cases. A gunshot is a pretty unambigu-
ous sound—but what about the sorts of crimes that we might imagine
a street-level surveillance camera capturing? Can software identify a
mugging in progress, for instance, with the same precision with which
it can identify a gunshot? If this book has insisted on anything, it's that
such problems—as difficult as they may seem—are more solvable than
they might appear at first glance.

THE FULL-TRANSPARENCY APPROACH

Thus far, I've been arguing that the same technology that's commonly
perceived as threatening privacy can also help us to develop advanced,
privacy-protecting controls. But there's also a simpler approach on
option: call it the full-transparency approach. Under this paradigm,
we would demand transparency from public safety agencies through
laws that allow any person to request data that they collected. In
general, transparency is a powerful concept, and it's likely one of the
best tools to make sure that agencies that can gather data about us are
held accountable.

In practice, though, this approach gets complicated. Forcing governments to open up access to their data could deter abuse of surveillance
powers. Or it could democratize surveillance, eroding privacy even further and faster.

Take the following example: under the *Public Records Act* of 1972,
the State of Washington mandates that government agencies cannot
deny requests for records if the requester is anonymous or the request is
too broad, nor can they deny requests to protect an individual's privacy.
Instead, they must redact only the details that are deemed sensitive
and release the rest. The law errs on the side of public transparency,
but while this sounds reasonable in the abstract, in practice it creates a
nearly unworkable set of constraints.

Beginning in 2014, Tim Clemans began sending release requests to
agencies across the State of Washington, with a focus on police body
camera footage. A spokesperson for the Seattle Police Department lik-
ened the requests to a distributed denial of service attack, a common
cyber-attack where hackers overload a website with traffic that causes

the site to crash. Eventually, Clemans created a bot capable of sending thousands of requests each day, overwhelming the ability of agencies to fulfill them. Body camera video can contain a wide range of private information that often needs to be redacted, from faces to names, addresses, and evidence of various types of crimes and victimization. Artificial intelligence isn't yet up to the task of automatic redaction, so agencies must do it manually. That can involve a frame-by-frame process, in which a human operator manually reviews and redacts the thirty images contained in every one second of video.

Paradoxically, the transparency required by the *Public Records Act* has created a practical impossibility for many agencies in the state. They have responded by not implementing body camera programs— reducing the transparency of policing. Even the laws intended to increase transparency to avoid abuse are themselves subject to the risk of abuse.

If you had begun this chapter hopeful that there would be a silver bullet, a single approach that would answer the challenges of balancing privacy and public safety, I am sorry to disappoint. Technology is moving much faster than regulations can keep up, and there is no one right answer to solve the problem. Even the answers that appear logical, like the Washington *Public Records Act*, may prove untenable in the face of the deluge of data and the challenges in meeting competing regulatory requirements, such as the directive to "release everything but redact all private information."

Technological tools, such as advanced AI redaction capabilities, may ultimately solve the cost and logistics problems before legislative efforts resolve them. Given the difficulty in predicting how technology will develop and how legislation will interact with it, regulatory frameworks will need to be agile and responsive.

I would argue that the erosion of privacy will have at least one beneficial effect: it will force governments to either enforce laws fairly or face a serious crisis of legitimacy. And assuming that they do not want to face such a crisis, governments will be pushed to change laws that they are unable to enforce fairly.

More broadly, we must face the fact that the world is ever more connected. As our social relationships move online, our personal data becomes part of a global information archive of unimaginable complexity. This same global internet is where terrorists recruit, communicate, and coordinate their activities. It is where we store our money, whether in digital bank accounts or online investment accounts. It is where we keep our most private memories and photos and conduct some of our most intimate conversations. It is where people express their cries for help or post their warnings of an impending violent outburst. As more of our lives shift online, public safety resources must move online as well. Whether they are fighting child sex crimes, terrorism, or cyber-theft, we could not exclude law enforcement from the online world even if we wanted to.

What used to be private thoughts or relationships are now digital trails of transactions, conversations, images, and videos. It will be difficult to fine-tune regulation of law enforcement's access to information. *Look here, but not there. Ignore this, but not that.* It can begin to feel like we are sticking our fingers into the holes of a giant dam of information that is bursting all around us.

Mark Zuckerberg, the CEO of Facebook, spoke for many leaders in government and technology when he suggested that "privacy is dead." I'm not ready to believe that just yet, and I still value keeping certain information away from others' prying eyes. But even if it's not dead, privacy is not what it once was. Our laws and our governments need to keep up.

ENGINE OF VIOLENCE:
ENDING THE WAR ON DRUGS

I N THE PREVIOUS chapter, I wrote that the erosion of privacy will force a choice on governments: they can enforce laws fairly or face increasingly difficult crises of legitimacy. The erosion of privacy also makes unequal enforcement of the laws more blatantly visible to the public. I think the best way to illustrate the point is by taking a closer look at an embattled policy: the American government's war on drugs.

The war on drugs is a powerful driver of killing in the United States today, creating an ecosystem that perpetuates violence and funds violent gangs by way of a shadow economy. If we are going to end killing, we should take a hard look at policies that perpetuate violence.

Let me lay my cards on the table: the racially and socially biased enforcement of some of our harshest laws is an injustice none of us should blithely accept. But that biased enforcement is inherent to the war on drugs as it has been carried out for decades.

If a student at Harvard uses cocaine, the general assumption is that he was a good person who made a bad decision. If he's caught, he'll likely get a slap on the wrist. Even more likely, he won't get caught in the first place, because campus police aren't generally assigned to parade through the dorms on high alert for cocaine. In the housing projects of Chicago or Miami, it's another story. If a kid gets caught with cocaine there, he's likely going to jail. His life will be put on a tragic trajectory of incarceration, criminality, heightened surveillance, and limited opportunities that can be impossible to alter.

By now, it's conventional wisdom that the war on drugs targets kids in communities of color and in low-income communities, far more than whiter and more affluent communities. But think about what this means: if the US government took seriously its commitment to equally and fairly enforce the laws on the books, narcotics officers would target places like Ivy League dorms, Silicon Valley offices, and Hollywood studios with the same zeal they bring to inner-city street corners.

That doesn't happen. Instead, drug laws are enforced with the harshest consequences on the least influential parts of society. If we fairly and truly enforced the drug laws across society, we would incarcerate a much larger segment of the population, including much of its leadership.

This is where privacy and legitimacy enter the picture. In the prior century, the government could offer a plausible-sounding excuse for unequal enforcement of drug laws. In places like inner cities, it could claim, drug use was relatively visible to law enforcement; it happened on street corners or in well-known crack houses. It was harder to know what went on inside a Harvard dorm than on the street corner of a city. Respectable places are respectable because they don't look from the outside like places where the law is being broken.

This has always been more of an excuse than a justification. But one upside of the fact that privacy has changed is that this excuse is no longer serious or credible. As more and more interpersonal communication moves online, the state could, if it chose, "see" into the Harvard dorm just as easily as it can see into an inner-city neighborhood. And as worrisome and invasiveness as that can seem, one benefit is that the state can no longer pretend to be unaware of crimes it doesn't bother to prosecute.

When the government can't plead ignorance, it has one of two options. The first is to admit to unequal enforcement of the laws—to say, "Yes, we're aware of what's going on in the Harvard dorm, but we're not interested in doing anything about it." I remain at least marginally hopeful that no democratic government could hold that line for very long.

Which brings us to the second option: scrap laws that we can't or won't enforce fairly. After all, it's not law enforcement's job to pick which laws to enforce. We want elected legislatures making laws, not police departments. And in a world where governments face increased pressure to enforce their laws fairly, the conversation about laws and lawmaking will change. The question won't be, "Who should bear the unfair brunt of the war on drugs?" It will be, "What laws could we support, if the government really did enforce them equally on all of us?"

And if we consider the question honestly, there's no way we could tolerate the war on drugs under that standard. More broadly, the changing nature of privacy could mean that there are new laws that are needed and old laws that need scrapping.

THIS BOOK came from the conviction that a huge amount of the killing in contemporary society is preventable, either through technological or policy change. The drug war is a prime example of the latter. Far from reducing violence, America's drug laws exacerbate violence in two key ways.

First, the war on drugs creates a shadow economy that is beyond the rule of law, leaving participants to enforce their rights and contracts themselves. That phenomenon is commonly called "street justice."

Second, the unfair overenforcement of these laws in communities of color undermines faith in the police as a fair and just protector of the community. When people lose faith in the government to protect them fairly and justly, they are more likely to take matters into their own hands, which also increases the rate of preventable violence.

In his book *The Better Angels of Our Nature*, Steven Pinker shows that the overall rate of violence has dropped dramatically as societies transitioned from ungoverned clans to modern states with a functioning public safety and justice system. His research shows that the risk of death from violence has dropped by a factor of five hundred times over the past thousand years, as modern government arose and took on the responsibility to protect its citizens and provide a framework to enforce their rights and resolve their conflicts without resorting to violence.

Pinker argues, convincingly, that inner cities have significantly higher rates of violence because key aspects of those communities are similar to "stateless" societies. In particular, inner-city economies depend on the illicit drug trade, a market in which participants cannot rely upon the state to protect and enforce their rights. As a result, much as medieval Europeans turned to an "honor code," or settlers in the Wild West resolved their differences with revolvers, residents of the "stateless" parts of the United States are more likely to protect their interests through violent means, filling a void created by the absence of reliable law enforcement. No drug dealer is about to call the police if a customer doesn't pay, so they take matters into their own well-armed hands.

Drug gangs and medieval knights would seem to have nothing in common—until you look at what motivates their behavior. In the context of their environment, violent behavior is a rational choice for protecting their interests. Without a functioning state that protects those interests, violence—particularly as a pre-emptive choice—is a logical, even if illegal and immoral, choice. The honor code of modern street gangs is not dissimilar, then, from the honor code of medieval Europe: enforce your rights with swift and severe violence upon those who have threatened your interests, before they have the chance to do the same.

Pinker further points out that reductions in violent behavior are predicated on members of society delegating the protection of their interests to the state. But this delegation only works if the members of that society believe that the state will fairly and impartially protect their interests. If they believe that the state is random or capricious, or worse, that the state actively targets them because of their race or other factors beyond their control, they will no longer delegate the protection of their interests to the state. They'll revert to violence to protect their interests.

Consider the current dynamic between police agencies and many communities. The issue isn't whether law enforcement officers are personally racist—certainly, the majority are not. The issue is that our laws, especially our drug laws, are enforced in a racially disparate way. But even if you want to dispute that fact, it's even harder to dispute the fact that the enforcement of drug laws is widely *perceived* as biased. This perception, in turn, drives members of some communities to lose faith in law enforcement and to refuse to delegate the use of force to the state. In this context, perception is reality. Communities that no longer trust the police to protect them will take matters into their own hands. Building community trust is more than a feel-good initiative; it's the foundation required for a community that renounces vigilantism and violent protection of self-interest.

Beyond driving violence and leaving vast swaths of the country as partial exceptions to the hopeful trends that Pinker describes, the war on drugs also has other harmful consequences. It contributes to the United States' status as home to the largest prison population on Earth—home to almost one-quarter of the world's prisoners, and nearly twice as many as China. It fuels an illicit economy, in which violence becomes the logical option for dispute resolution. It undermines faith in law enforcement. And yet, despite the heavy price tag, we have little to show for it all in terms of declining drug use. Consider figures 7 to 10, which lay our failure out in graphic terms.

First, look at the explosion of the American prison population (figure 7).

Notice that it begins to increase exponentially in the early 1970s, when the Nixon administration first declared the war on drugs.

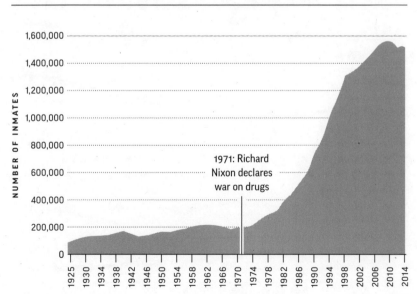

FIGURE 7: U.S. STATE AND FEDERAL PRISON POPULATION · 1925–2014

But if we are incarcerating so many drug users and dealers, are we at least reducing drug use? If we are spending billions on incarceration, are we at least achieving the stated policy goals of the war on drugs? Again, the numbers speak for themselves.

Consider figures 8 to 10, all from the National Institute on Drug Abuse.

What do these charts tell us? First, they tell us that illicit drug use is growing across the entire population (figure 8). Specifically, it's even growing among more "conservative" demographics such as adults aged fifty to sixty-four (figure 9). Even as we have been executing the war on drugs, drug use is flat or increasing.

And yet there's one substance whose use is dramatically in decline: tobacco (figure 10). We have prevented countless kids from taking up the smoking habit, prevented countless cases of cancer, and saved countless lives—all without making tobacco illegal or creating illicit markets. We've regulated it, we've advertised against it, we've built social stigma against it—but we haven't ever called for an enforcement-driven "war on tobacco."

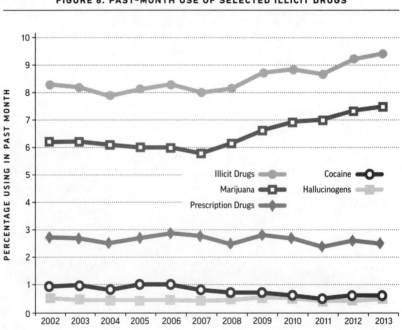

FIGURE 8: PAST-MONTH USE OF SELECTED ILLICIT DRUGS

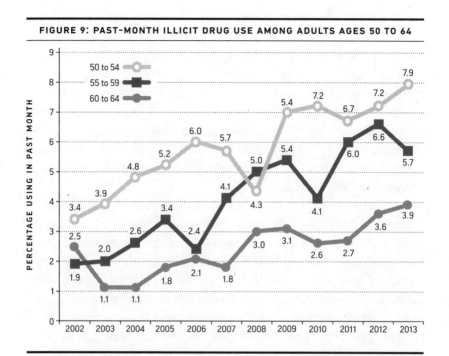

FIGURE 9: PAST-MONTH ILLICIT DRUG USE AMONG ADULTS AGES 50 TO 64

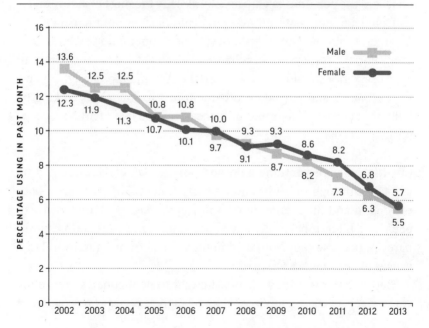

FIGURE 10: PAST-MONTH CIGARETTE USE AMONG YOUTHS AGED 12 TO 17

We have a proven strategy for driving down the use of a harmful substance, and it looks far more like our anti-tobacco policies than the war on drugs. We've made illicit drugs illegal, and we've fiercely enforced laws against their use—all without putting a dent in that use. On the contrary, we've created a criminal market that fuels gangs and violence.

I can't put it better than Richard Branson, founder of the Virgin Group and a member of the Global Commission on Drug Policy: "As an investment, the war on drugs has failed to deliver any returns. If it were a business, it would have been shut down a long time ago. This is not what success looks like."

Let me return to my point on biased enforcement. According to the *Washington Post* in 2015, the United States incarcerates 716 out of every 100,000 people, or a little under 1 percent. But if we consider figure 8 and assume that at least 9 percent of the population are using illicit drugs, we would have to theoretically increase the population of our prisons nine-fold if we fairly enforced the drug laws across the entire population. In other words, our system only functions today because of the institutionally biased enforcement of the laws we have on the books—because most drug users escape the harsh penalties set by our laws, but those on whom the penalties fall are usually already the most disadvantaged.

Could a nation really afford—politically, socially, or financially—to incarcerate more than a tenth of its entire population? Almost certainly not: the system would collapse. We can't afford to incarcerate all drug users, nor should we want to. So the system limps along, depending on biased and partial enforcement just to stay alive, costing us billions in wasted dollars and millions of wasted lives, without accomplishing its goals.

I DON'T make this argument because I'm in favor of widespread drug use. Quite the opposite. My family has experienced the scourges of both the opioid and the methamphetamine epidemic. My family has also been hard hit by tobacco use and abuse. I never met my paternal grandparents, because they both died in their fifties from cigarette-related causes (emphysema and stroke).

Because of my family history, I ought to be inclined to more fervently back a "war on tobacco" than to support a "war on drugs." But as a parent, what I care about are effective strategies that reduce the negative consequences of using drugs or tobacco overall. And I am even more focused on eliminating the policies that incubate violence. Every available fact tells us that our anti-tobacco strategy has outperformed our anti-drug strategy: reducing usage and avoiding the heavy costs of violence, incarceration, and mistrust of police.

Effective strategies need not emphasize toughness, a point that is borne out by evidence from around the globe. A 2014 study of anti-drug programs of different countries by the Home Office in the United Kingdom concluded, "There is no apparent correlation between the 'toughness' of a country's approach and the prevalence of adult drug use." In Portugal, a breakthrough model has shown how a country can reduce drug use, using the tools of public health while avoiding an over-emphasis on law enforcement. Portugal decriminalized many drugs in 2001 and then built programs to help addicts. Over that decade and a half period, drug use and related effects are going down. Portugal's strategy appears to be outperforming the United States'.

These are controversial views for someone in my position: my company's primary customers in law enforcement disagree with the decriminalization of illicit drugs. In my discussions with law enforcement officers, I've often heard compelling argument for keeping the laws against drugs: these laws help law enforcement officials investigate and remove violent offenders from the streets.

I take that argument seriously, because being a police officer in the United States is a difficult and frustrating job. The constitutional rights meant to protect us all can make it difficult for law enforcement to arrest and remove those who threaten our safety. Most officers have a sincere belief that they need every tool—including the laws on the books prohibiting illicit drug use—to effectively investigate and arrest violent offenders. For them, without probable cause related to drug infractions, it could be much harder to get violent offenders off the street.

I believe this is true, in a practical sense. A police officer squares off against violent, dangerous people who don't have to play by the rules. Police, on the other hand, are constrained by a complex set of rules that govern all of their actions. Add to that the frustration of seeing so many of these offenders get released by the courts because of some legal technicality (or the lack of sufficient court and prison capacity), and you can understand why police aren't ready to give up on drug laws.

But if we look at violence over the long term, I strongly believe that Pinker is right: our drug laws are creating a stateless economy that incubates violence. If we removed the laws that criminalize the illicit drug trade, police officers may have more trouble getting violent offenders off the streets in the short term. But in the long run, we can reasonably expect to create a system that produces far fewer violent offenders.

Don't get me wrong: I don't think we should let violent offenders off the hook for their crimes. But if we know we have created a flawed system that leads to violence, we should address that system at its roots—and not tinker at its margins. Our world would be better off with fewer violent criminals, period, not with laws that create those criminals and then force police to deal with them.

As someone who works closely with police around the country, I also object to the idea of asking police to put their lives on the line every day to fight a war with no clear path to victory. We should not ask police to risk their lives enforcing laws that, by all accounts, it appears we don't, deep down, want them to enforce. Even the most ardent law-and-order conservative is likely to have a relative who has a drug problem, and when it comes to our families, we have no desire to see them go to jail. And remember: if the United States really walked the walk on drug enforcement, we'd have more than a tenth of the population in prison.

So to sum up: we're in a war we have no real intention of winning, and with no clear plan to win. In a war like that, you rethink your strategy. You don't keep sacrificing lives—you change your approach. And when you're winning a similar conflict with a different strategy—in this case, the war on tobacco—you borrow the best of that playbook.

To emphasize again: I am not advocating or promoting drug use or abuse. Too often, this debate is characterized in extremes: you either want to send every drug user to jail, or you want to celebrate the mind-opening possibilities of free-flowing drugs in our society. I reject that false dichotomy. You can accept that drugs destroy lives, without therefore believing that the drug war is the best way to reduce drug use. We should find a middle path, one where we gain the benefits of decriminalization with the long-term reductions of violence in black market economies. That path would also lead us to create programs to drive down drug use and its attendant negative effects.

There are voluminous books dedicated to analyzing the criminalization of illicit drugs. My intention here is not to dive into the complexities of drug policy, but rather to use this area to illustrate the challenges of applying old laws to a new world where privacy cannot be what it once was. I'm also motivated to tackle this issue because of its deep connection to the overriding goal of this book: to reduce violence in our communities.

These are complex issues, and there's no single answer. But I'm motivated by a vision of a world in which police officers are truly public safety officers. To achieve this vision, we need laws that allow them to advance society's best interests, tools that avoid creating lasting harm, and training that helps them to become their communities' champions. All of this begins with balanced, effective laws. To create this ideal, we need to make sure the laws we are asking to be enforced strike the right balance between protecting individual rights and maintaining a peaceful, safe, and functional society.

We owe it to the men and women in blue to take a hard look at what laws we actually want them to enforce, fairly and consistently, for all of us. And we owe it to our citizens to have laws that we fairly enforce and adjudicate. If a law can't live up to that standard, we must consider whether it is past its shelf life.

I'll come back to the point I made at the outset: a democracy should only tolerate laws that are enforced for everyone. We're not willing to enforce our drugs laws for everyone. We've seen how the existing laws drive increased violence by creating an illicit economy, one in which participants turn to violence to protect interests that cannot be protected by police and courts. Those laws are ineffective and even actively harmful to public safety. So we should simply, and responsibly, end them.

CONCLUSION
A PATH FORWARD

E CAN, AND must, think about violence, use of force, and the security of society in a new way. We can, and must, move past the notion that killing is the only means of securing the peace. And we must shift toward a new paradigm, one in which non-lethal force is the default solution.

I am confident that one day we will see non-lethal options as the only rational choice. I hope I've shown how our notions of acceptable violence have changed over time and how they continue to change. And I hope you've come away with some appreciation for the remarkable new technologies that will, one day soon, make the end of killing a real possibility.

But I hope I haven't given you the wrong idea: the path ahead will not be easy. It won't be simple or without difficulties. Working on the front lines of technology, I've learned first-hand that the forces arrayed against social and technical progress are powerful and should not be underestimated. I think that these forces draw their effectiveness from a basic aspect of human psychology—what I have referred to as the phenomenon of the new versus the now. We humans are naturally risk-averse animals. When faced with the possibility of a radical change—a new tool, a new policy, or a new way of organizing our societies—we don't compare it to the status quo, dispassionately weighing each for pros and cons. We instinctively compare it with perfection, focusing heavily on the flaws in what is new rather than the improvements over what exists today. In other words, as soon as we discover a problem with the new—and there are always problems, because humans are incapable of creating perfection—we tend to discard or disregard it.

This phenomenon manifests itself in media coverage that generates controversy and gains public attention by fixating on the downsides of new approaches: the prototype that fails, the pilot program that goes over budget, the experiment with ambiguous results. It also manifests itself in entrenched bureaucracies, in both the private and public sectors, made up of people who benefit from the status quo and might stand to lose if it changes. It manifests itself in regulations that protect current stakeholders rather than enabling change in the public interest. And it manifests itself in the way we think of risk.

In this book, I've outlined some radical possibilities for the future of non-lethal force—possibilities that may seem as "out there" as the TASER seemed a few decades ago or gunpowder must have seemed to armored and mounted knights. It's natural to worry about the risks inherent in developing and implementing the technologies I've talked about. But if there's one thing I hope you'll take with you, it's that there is always risk on both sides. The risks of the status quo become so familiar to us that they start to seem invisible. And while there are risks in some of the approaches outlined in this book, we must consider the risks of not taking them—the risks of remaining in a world where the default is to build peace by dealing out death.

I'm passionate about getting risk assessment right because I've already lived through the process I've described. If you can think of a line of criticism directed at a non-lethal weapon, I've been on the receiving end of it. All of those criticisms were directed at the TASER weapon. It was a toy from The Sharper Image. It was a weapon for wimps. It was forced on cops by the suits at city council. It would trick cops into becoming dependent on them—they'd forget how and when and whether to fire their guns. After TASERS gained widespread adoption, the narrative focused on the dangers of the weapon, with the unrealistic expectation that it would make high-risk situations risk-free. It was a "kinder and gentler" weapon, when the whole point of weapons is to be unkind and ungentle. And of course, when TASER weapons were misused, the technology, more than the misuse itself, was suspect.

In the abstract, each of the critiques can seem reasonable. However, we do not live in an abstract reality; we live in a reality marred by great imperfections. Given the state of the world as it is, we must measure the new against the status quo, not in abstract isolation, which leaves us comparing to perfection. Put the status quo and the future on the scales and weigh them fairly. And eventually, if these sorts of arguments hit home with just a few influential leaders, then the future is on the way to becoming a status quo of its own. The path of almost all transformative technology goes from resistance to reluctance to acceptance to dependence. That's the path we can expect from all of the technology outlined in this book.

As Peter Diamandis put it, "The day before something is truly a breakthrough, it's a crazy idea." Today, an end to killing sounds like a crazy idea. Totally non-lethal policing sounds like a crazy idea. A military that conducts operations without bloodshed sounds like a crazy idea. Activists and tech companies cooperating with police and military institutions sounds like a crazy idea. A world free of school shootings sounds like a crazy idea.

Today, it may sound laughable. Tomorrow, it will sound as if it had been inevitable.

ACKNOWLEDGMENTS

WRITING A BOOK is a bit like releasing a gas in a vacuum: it expands to fill the empty space, or in this case, the empty time. So, I'd like to start out by extending a heartfelt "thank you" to my family, who felt this project in the empty space left when I wasn't present these past few years. Nights, weekends, and vacations became the chunks of time where I would turn my attention away from my day job to my passion project, a goal to start a movement to bring an end to killing. In many ways, the price was more heavily paid by my loved ones, who felt the strain of my absence (and the stress of my presence) without feeling the reward of creation.

While I thought writing a book would involve me, a laptop, and a remote cabin in some idyllic setting, it turned out it took a team of people to bring this project to life. Trena White, Rony Ganon, Amanda Lewis, Tilman Lewis, and the entire team at Page Two have been invaluable in providing all the services of a publisher, without all the old-school baggage. Casper, my friendly ghostwriter, helped transform these ideas into well-formed concepts and made a great intellectual sparring partner as we beat these ideas against the rocks to see which would survive.

This book would not have been possible without the more than one thousand people who have joined with me to create TASER International, then to transform it into Axon Enterprise. That starts with Jack Cover, who taught me how to build advanced TASER weapons in his garage, and his wife Ginny, who allowed us to hijack large portions of her home for months at a time as we got this effort off the ground. Without my brother Tom setting up operations from bank accounts to office space, we might still be in a garage somewhere. And, without Hans Marrero, the company may have died as a failed consumer product experiment, without ever making the transition into the world of public safety.

There have been countless experts in law enforcement and military affairs, who welcomed a novice outsider and took the time to educate me and guide me along the way. From Sid Heal, one of the early evangelists of non-lethal weapons, to Greg Meyer, the LAPD sergeant (later Captain) who led the early charge of TASER into policing, your help was invaluable. Chuck Ramsey, Jeff Halstead, and Darrel Stephens have been among the hundreds of police chiefs who took me under their wings along the way and gave critical feedback to help guide this project. Similarly, Generals Zinni, Petraeus, and Bedard together with Colonels Buran and Fenton and Lieutenant Colonel Dave Grossman provided great insights into the challenges—and great promise!—of introducing non-lethality into military institutions. Finally, Scott Greenwood and Barry Friedman helped immeasurably in teaching me how to listen constructively and make ideas stronger through diversity of input.

Finally, to the men and women of law enforcement, military, and public safety: I am amazed by your willingness to get up every day, strap on a gun and body armor, live with the uncertainty you may not come home, and put yourself at risk to protect people you do not know—and who may be more likely to criticize you than to thank you. Thank you for accepting me into your community, and for giving me the latitude to challenge traditional ways of doing business. Your trust is deeply appreciated. We are at the beginning of a great journey, and I look forward with excitement to seeing where it leads us.

NOTES

INTRODUCTION

Almost 40,000 people a year are killed... Sarah Mervosh, "Nearly 40,000 People Died from Guns in U.S. Last Year, Highest in 50 Years," *New York Times,* December 18, 2018. (Note: about half of those firearm deaths in the United States are suicides.)

Companies are currently developing the next generation of lethal and non-lethal tools... "Non-lethal," "less-lethal," and "less-than-lethal" are three terms for the same concept. A significant amount of confusion and controversy have surrounded the question of what to call weapons that are designed not to kill. The terms "non-lethal," "less-lethal," and "less-than-lethal" are all terms for the exact same thing: weapons that are designed to deter or stop a threat without killing the target. Sometimes people misperceive these terms to describe varying levels of danger, as if a less-lethal weapon were in a more dangerous category than a non-lethal weapon. This is a false dichotomy. For simplicity, I use "non-lethal" throughout this book, as I believe it is the simplest, most widespread label. It remains the term of choice in both academia and the military. See www.EndOfKilling.com for more on the distinction between these terms.

... we will need clear and ambitious goals... A skeptic could fairly point out that these objectives would benefit my company and me, personally. This is true, much as the early innovators in the computer industry stood to benefit from a vision of a computer in every household.

I acknowledge the alignment of my interests to these goals, but that should not be the end of the discussion. I would ask for an open mind as you evaluate whether these goals stand on their own as progress toward a better world than the one that exists today.

1: WEAPONS, PAST AND PRESENT

On the savanna of Senegal in 2007... J.D. Pruetz et al., "New Evidence on the Tool-Assisted Hunting Exhibited by Chimpanzees (*Pan troglodytes verus*) in a Savannah Habitat at Fongoli, Sénégal," *Royal Society Open Science* 2, no. 4 (April 1, 2015), http://doi.org/10.1098/rsos.140507.

The atlatl, whose invention dates... C.D. Howard, "The Atlatl: Function and Performance," *American Antiquity* 39 (1974), 102–104.

Homer's Iliad *offers a vivid description*... Homer, *Iliad*, trans. A.T. Murray (Cambridge, MA: Harvard University Press, 1924), book 3, Perseus Digital Library at Tufts, http://perseus .uchicago.edu/perseus-cgi/citequery3.pl?dbname=GreekTexts&Query=Hom.%20Il.&getid=2.

... by 1132, soldiers of the Song dynasty had used the first proto-firearms in battle... Tonio Andrade, *The Gunpowder Age: China, Military Innovation, and the Rise of the West in World History* (Princeton, NJ: Princeton University Press, 2016).

... the invention of the first fully automated firearm in 1884... "How the Machine Gun Changed Combat During World War I," Norwich University Online, https://online.norwich.edu/ academic-programs/masters/military-history/resources/infographics/how-the-machine- gun-changed-combat-during-world-war-1.

... Foucault showed how the public execution was choreographed... Michel Foucault, *Discipline and Punish: The Birth of the Prison*, trans. Alan Sheridan (New York: Vintage, 1995).

Consider another execution, which took place in 2018... Tracy Connor, "Doyle Lee Hamm Wished for Death during Botched Execution, Report Says," NBC News, March 5, 2018, https:// www.nbcnews.com/storyline/lethal-injection/doyle-lee-hamm-wished-death-during-botched- execution-report-says-n853706.

2: THE NEW VS. THE NOW

... an average of 3,287 deaths happen on the road each day... "Road Safety Facts," Association for Safe International Road Travel, https://www.asirt.org/safe-travel/road-safety-facts/.

In 2013, Elon Musk wrote a convincing piece... Elon Musk, "The Mission of Tesla" (blog), Tesla, November 18, 2013, https://www.tesla.com/blog/mission-tesla.

Did you know that automotive airbags were invented as early as 1951?... Kimberlea Buczeke, "The Evolution of Airbags," *RepairPal Blog*, May 18, 2017, https://repairpal.com/blog/ history-of-airbags.

3: WHY WE NEED TO STOP KILLING

The former police commissioner of Philadelphia, Chuck Ramsey... Author interview with Chuck Ramsey, July 6, 2018.

... Hans Marrero, a former Marine, said to me... Author interview with Hans Marrero, June 14, 2018.

Dr. Maguen wanted to see... "The Wounds of the Drone Warrior," *New York Times Magazine*, June 13, 2018, https://www.nytimes.com/2018/06/13/magazine/veterans-ptsd-drone-warrior- wounds.html.

As one such study put it, "Prior research has found that many officers involved in shootings..." "Police Responses to Officer-Involved Shootings," *NIJ Journal* no. 253, National Institute of Justice, https://nij.gov/journals/253/pages/responses.aspx.

... as documented in 2011 in the Journal of Psychiatric Research... Irina Komarovskaya et al., "The Impact of Killing and Injuring Others on Mental Health Symptoms among Police Officers," *Journal of Psychiatric Research* 45, no. 10 (October 2011), 1332–1336, https://www.ncbi.nlm.nih. gov/pmc/articles/PMC3974970/.

Dr. Gregory Elder, a Defense Intelligence Agency and Central Intelligence Agency researcher, illustrates how... Gregory Elder, "Intelligence in War: It Can Be Decisive," *Studies in Intelligence* 50, no. 2 (2006), Unclassified Edition, Center for the Study of Intelligence, https://www.cia.gov/library/center-for-the-study-of-intelligence/csi-publications/csi-studies/studies/vol50no2/html_files/Intelligence_War_2.htm.

As one Taliban official commented to the New York Times *in early 2019...* Taimoor Shah and Fahim Abed, "Airstrikes in Taliban Area Kill 29 Afghans Despite Peace Talks," *New York Times*, January 25, 2019, https://www.nytimes.com/2019/01/25/world/asia/afghanistan-airstrikes-taliban.html.

4: THE TASER STORY

Over the last twenty years, TASER weapons... J. Brewer and M. Kroll, "Field Statistics Overview," in M. Kroll and J. Ho, eds., *TASER Conducted Electrical Weapons: Physiology, Pathology, and Law* (New York: Springer-Kluwer, 2009).

Dr. Laurence Miller, a clinical and police psychologist... Laurence Miller, "Suicide by Cop: Causes, Reactions, and Practical Intervention Strategies," *International Journal of Emergency Mental Health* 8, no. 3 (2006), 165-174, https://www.psychceu.com/miller/Miller_Suicide_by_Cop.pdf.

5: VIOLENCE IS LIKE A VIRUS: PUBLIC SAFETY

Several studies have shown that, with respect to sexual assaults... "Victims and Perpetrators," National Institute of Justice, last modified October 26, 2010, https://www.nij.gov/topics/crime/rape-sexual-violence/pages/victims-perpetrators.aspx.

... being a member of a certain social network increased the odds of being a victim of homicide by 900 percent... Shankar Vedantam, "Research May Give Potential Homicide Victims a Heads Up," NPR, October 9, 2014, https://www.npr.org/2014/10/09/354754588/research-may-give-potential-homicide-victims-a-heads-up.

Bratton himself agreed: "CompStat was misused in the 21st century..." William Bratton, "Cops Count, Police Matter: Preventing Crime and Disorder in the 21st Century," report for The Heritage Foundation, September 21, 2017, https://www.heritage.org/crime-and-justice/report/cops-count-police-matter-preventing-crime-and-disorder-the-21st-century.

The City of Chicago already deploys a network... Scott Goldfine, "32K Surveillance Cameras Aim to Keep Chicago Safe," *Campus Safety*, October 31, 2018, https://www.campussafetymagazine.com/technology/surveillance-cameras-keeping-chicago-safe/.

"Spotlight uses natural language processing..." "Ashton Kutcher Talks to *48 Hours* about His Fight to End Child Sex Trafficking," A Plus, April 1, 2018, https://aplus.com/a/ashton-kutcher-child-sex-trafficking-thorn-48-hours.

Friedman writes of his methodology, "In this fraught space, there are three questions..." Maria Ponomarenko and Barry Friedman, Benefit-Cost Analysis of Public Safety: Facing the Methodological Challenges," *Journal of Benefit-Cost Analysis* 8, no. 3 (Fall 2017), 305-329, https://www.cambridge.org/core/journals/journal-of-benefit-cost-analysis/article/benefitcost-analysis-of-public-safety-facing-the-methodological-challenges/769C3A503BBCF521AF40BDFB61F17B02/core-reader.

John Webster from the University of Illinois, Chicago, tracked the activities of the patrol division of an agency... John A. Webster, "Police Task and Time Study," *Journal of Criminal Law and Criminology* 61, no. 1 (1970), https://scholarlycommons.law.northwestern.edu/jclc/vol61/iss1/9/. Results are from an internal study of a national police force. The results are not public, but the high-level result was conveyed in a private meeting with police leadership.

... among the most effective measures for driving down violent crimes is to put more officers on the streets... Rina Palta, "Study: Hiring More Police Cuts Violent Crime," *The Latest* (blog), November 9, 2012, https://www.scpr.org/blogs/news/2012/11/09/11019/study-hiring-more-police-cuts-violent-crime/.

In a survey of twelve thousand police chiefs and command staff, Nuance Communications found... Mark Geremia, "Survey Finds Inefficient Documentation Processes Cost Officers Time," *What's Next* (blog), Nuance, January 30, 2018, https://whatsnext.nuance.com/office-productivity/mark-geremia-paperwork-in-police-work-survey/.

In 2016, Dallas police officers used a robot rigged with explosives... This is drawn from both an interview the author conducted with Scott Greenwood as well as the news coverage of this particular incident.

As one American Civil Liberties Union leader, Scott Greenwood, put it to me... Author interview with Scott Greenwood, June 27, 2018.

6: EXTENDING THE TREND: MILITARY

A single detonation in Hiroshima... Kirk Spitzer, "Hydrogen Bomb vs. Atomic Bomb: What's the Difference?," WLTX, September 4, 2017, https://www.wltx.com/article/news/nation-world/hydrogen-bomb-vs-atomic-bomb-whats-the-difference/470619324.

A hundred thousand drones that cost $2,000 each... Colin Ritsick, "F-22 Raptor vs F-35 Lightning II," Military Machine, January 8, 2019, https://militarymachine.com/f-22-raptor-vs-f-35-lightning-ii/.

The head of product development at Uber recently predicted... James Hetherington, "Uber's Big Breakthrough: San Francisco to San Jose in Just 15 Minutes," *Newsweek*, May 5, 2018, https://www.newsweek.com/ubers-big-breakthrough-san-francisco-san-jose-just-15-minutes-916782.

Many technologists point out that we do not yet even understand how the brain stores information... The current techniques to map the brain involve painstakingly chopping the brain into slices and observing them under a microscope. See Shelly Fan, "Amazing New Brain Map of Every Synapse Points to the Roots of Thinking," Singularity Hub, August 14, 2018, https://singularityhub.com/2018/08/14/amazing-map-of-every-synapse-in-the-mouse-brain-points-to-the-roots-of-thinking/.

BRAIN (Brain Research through Advancing Innovative Neurotechnologies) Initiative... "The BRAIN Initiative: Brain Research through Advancing Innovative Neurotechnologies," Obama White House Archives, last modified February 2015, https://obamawhitehouse.archives.gov/BRAIN.

The computers used to process the data from gene sequencing machines were doubling in performance every eighteen months... James M. Heather and Benjamin Chain, "The Sequence of Sequencers: The History of Sequencing DNA," *Genomics* 107, no. 1 (January 2016), 1–8, https://www.sciencedirect.com/science/article/pii/S0888754315300410.

And that is exactly what happened with the Human Genome Project... Lisa Gannett, "The Human Genome Project," *Stanford Encyclopedia of Philosophy* (Summer 2016), Edward N. Zalta (ed.), https://plato.stanford.edu/archives/sum2016/entries/human-genome/.

...if the computing power available to map the human brain continues to double about every eighteen months... Note that the narrow exponential performance improvements in Moore's Law relate specifically to the number of silicon transistors that can fit in a given unit area.

Moore's Law is reaching the limits of physics as the size of logic gates approach single-molecule scale. However, a host of new technologies seem poised to pick up where transistors hit their maximum performance, whether it be new 3D chip architectures, quantum computing, optical computing chips, carbon nanotubes, or other emerging technologies are likely to fill the void and continue the acceleration of computing. ("After Moore's Law," *Technology Quarterly, Economist,* March 12, 2016, https://www.economist.com/technology-quarterly/2016-03-12/after-moores-law.)

...in 2017, 17.5 million people chose to have cosmetic surgery... "2017 Plastic Surgery Statistics Report," American Society of Plastic Surgeons, https://www.plasticsurgery.org/documents/News/Statistics/2017/plastic-surgery-statistics-full-report-2017.pdf.

An additional 860,000 people had even more invasive surgeries... "AJRR Releases 2017 Annual Report on Hip and Knee Arthroplasty Data: Report Gives Most Comprehensive Picture to Date of U.S. Hip and Knee Replacement Patterns," American Joint Replacement Registry Newsroom (press release), November 3, 2017, http://www.ajrr.net/media-news/press-releases/500-ajrr-releases-2017-annual-report-on-hip-and-knee-arthroplasty-data.

7: A CULTURAL INSURGENCY

When the U.S. military is entrusted with responsibility for security in another country... David C. Gompert et al., *Underkill: Scalable Capabilities for Military Operations amid Populations* (Santa Monica, CA: RAND Corporation, 2009), https://www.rand.org/pubs/monographs/MG848.html.

The Department of Defense has a budget of over $700 billion... United States, Under Secretary of Defense (Comptroller), "DoD Budget Request: Defense Budget Materials—FY2019," https://comptroller.defense.gov/Budget-Materials/.

"Screw the Army," Zinni said... I've been blessed to have friends in all branches of the US Military, but for those who might be less familiar with it, Zinni's comments reflect a good-natured rivalry between the Army and Navy (of which the Marine Corps is technically one department). It also reflects the rivalry between the Marine Corps and, well, all other branches of the US Military. Though he was speaking candidly, this particular line was delivered with a smirk.

"For the first millisecond, it just felt like the skin was warming up..." Jurica Dujmovic, "Futuristic War Weapons Are Already Here: Lasers, Electromagnetic Rails and Microwave Rays," *MarketWatch,* April 14, 2015, https://www.marketwatch.com/story/futuristic-war-weapons-are-already-here-lasers-electromagnetic-rails-and-microwave-rays-2015-04-14.

..."a bit like touching a red-hot wire, but there is no heat..." "New Frontiers in Torture," CBS News, Political Animal (unsigned column), October 8, 2007, https://www.cbsnews.com/news/new-frontiers-in-torture/.

... *"the non-lethality of the ADS system could prove useful in a counterinsurgency operation..."*
David J. Trachtenberg, "An Opportunity Missed," American Enterprise Institute Center for
Defense Studies, August 30, 2010, https://web.archive.org/web/20141101231037/http://
www.defensestudies.org/cds/an-opportunity-missed/.

8: JUDICIOUS SURVEILLANCE

The morning of March 22, 2016, started out as a day like any other at Brussels Zaventem airport...
Henry Austin, "Brussels Attacks: Timeline of the Events in the Belgian Capital," NBC News,
March 22, 2018, https://www.nbcnews.com/storyline/brussels-attacks/brussels-attacks-
timeline-events-belgian-capital-n544221.

In 2015, agents from the Department of Homeland Security posed as passengers... Tom Costello
and Alex Johnson, "TSA Chief Out after Agents Fail 95 Percent of Airport Breach Tests," NBC
News, June 1, 2015, https://www.nbcnews.com/news/us-news/investigation-breaches-us-
airports-allowed-weapons-through-n367851.

In an astonishing moment, a Homeland Security official posing as a passenger... Justin Fishel,
Pierre Thomas, Mike Levine and Jack Date, "Exclusive: Undercover DHS Tests Find
Security Failures at US Airports," ABC News, June 1, 2015, https://abcnews.go.com/US/
exclusive-undercover-dhs-tests-find-widespread-security-failures/story?id=31434881.

... Synapse demonstrated 98.73 percent effectiveness... Author interview with Ian Cinnamon,
Synapse founder, February 14, 2018.

On February 5, 2016, Nikolas Cruz was pictured with guns on Instagram... Brett Murphy and
Maria Perez, "Florida School Shooting: Sheriff Got 18 Calls about Nikolas Cruz's Violence,
Threats, Guns," USA Today, February 23, 2018, https://www.usatoday.com/story/news/
nation-now/2018/02/23/florida-school-shooting-sheriff-got-18-calls-cruzs-violence-threats-
guns/366165002/.

9: JUSTICE MODERNIZED

Leading molecular biologists say a technique promoted by the nation's top law enforcement agency...
Gina Kolata, "Some Scientists Doubt the Value of 'Genetic Fingerprint' Evidence," *New York
Times*, January 29, 1990, https://www.nytimes.com/1990/01/29/us/some-scientists-doubt-
the-value-of-genetic-fingerprint-evidence.html.

Gary Leon Ridgway was arrested in 2001... William Booth, "A Long-Sought Break in Green
River Killings," *Washington Post*, December 2, 2001, https://www.washingtonpost.com/
archive/politics/2001/12/02/a-long-sought-break-in-green-river-killings/b30abcda-2b63-
4c5a-b060-1e8d646237e4/.

It wasn't until 1997 that DNA evidence established that Criner couldn't have been the rapist...
"Roy Criner," Innocence Project, August 15, 2000, https://www.innocenceproject.org/cases/
roy-criner/.

*... as the DNA example proves, the use of technologies to help us get at the truth in criminal justice
cases can save lives and advance the cause of justice...* People must continue to invest the time and
effort into furthering development of promising technologies. When I came across Jack Cover,
TASER technology was effectively dead. Two companies had tried, both had gone bankrupt,
and the only vestige of them left was a tiny two-person operation that was selling only a few

hundred devices per year. If we hadn't convinced Jack to give it another try and hadn't raised the money from friends and family, I believe the technology would have disappeared for good.

As with the first iteration of DNA, the early TASER devices had limitations. They had bugs. They were clunky. And that was almost enough to stop the idea in its tracks. But on the third attempt, in the third decade since invention, we finally got it "right enough" to scale a business that is supporting ongoing research and further progress. We have more work to do to make this technology a worthy substitute for lethal force, but that work is getting done—only because we have pushed through to prove a threshold of utility that has passed the barrier for general acceptance of the technology.

Trial by jury is subject to all kinds of irrational pressures ... Because I am in the business of selling TASER weapons to police, my company is frequently involved in litigation around police use of force. We need to understand more than just the science of TASER technology. We need to understand how human beings will react to the science. So we conduct practice presentations to jury focus groups. One consistent thing that amazes me is how little facts seem to drive juries' decisions, compared to emotional connections that take place between jurors and between the dominant juror (sometimes referred to as "the lion") and the litigators. The lion is a natural leader who tends to emerge to drive the jury's discussions and deliberations. That one juror's perspective is often the main determining factor in the outcome of a case. Even more troubling, the juries we've observed spend a great deal of time discussing what the attorneys and witnesses are wearing, or who seems "likeable"—often more than on the hard facts of the case at hand.

In 1969, South Africa banned jury trials altogether ... Corey Adwar, "Here's Why Nobody in South Africa Gets a Jury Trial including Oscar Pistorius," *Business Insider,* April 3, 2014, https://www.businessinsider.com/heres-why-oscar-pistorius-wont-get-a-jury-trial-2014-4.

... a 2007 Northwestern study found ... Bruce Spencer, "Estimating the Accuracy of Jury Verdicts," *Journal of Empirical Legal Studies* 4, no. 2 (July 2007), 305-329, https://doi.org/10.1111/j.1740-1461.2007.00090.x.

"A scientist has to make inferences about states of affairs that cannot be observed directly..." Jonathan Capehart, "'Hands Up, Don't Shoot' Was Built on a Lie," *Washington Post,* Opinion, March 16, 2015, https://www.washingtonpost.com/blogs/post-partisan/wp/2015/03/16/lesson-learned-from-the-shooting-of-michael-brown/.

... over 70 percent of the large city police departments in the United States issue body cameras ... Charlie Frago and Tony Marrero, "Body Cameras Now Standard Gear for Florida Cops—But Not in Tampa Bay," *Tampa Bay Times,* March 20, 2018, https://www.tampabay.com/news/publicsafety/Body-cameras-now-standard-gear-for-Florida-cops-But-not-in-Tampa-Bay_166350517.

The story of Sergeant Brandon Davis ... Quentin Hardy, "Taser's Latest Police Weapon: The Tiny Camera and the Cloud," *New York Times,* February 21, 2012, https://www.nytimes.com/2012/02/21/technology/tasers-latest-police-weapon-the-tiny-camera-and-the-cloud.html.

In 2013, California's Rialto Police Department ran one of the earliest tests of body-worn cameras ... Tony Farrar, *Self-Awareness to Being Watched and Socially-Desirable Behavior: A Field Experiment on the Effect of Body-Worn Cameras on Police Use-of-Force* (Washington, DC: National Police Foundation, March 2013), https://www.policefoundation.org/publication/self-awareness-to-being-watched-and-socially-desirable-behavior-a-field-experiment-on-the-effect-of-body-worn-cameras-on-police-use-of-force/.

A more recent 2016 study out of Cambridge University... "Use of Body-Worn Cameras Sees Complaints against Police 'Virtually Vanish', Study Finds," University of Cambridge Research (press release), September 29, 2016, https://www.cam.ac.uk/research/news/use-of-body-worn-cameras-sees-complaints-against-police-virtually-vanish-study-finds.

One study put the estimated annual cost of our current model of incarceration at a staggering one trillion dollars... Matt Ferner, "The Full Cost of Incarceration in the U.S. Is over $1 Trillion, Study Finds," *Huffington Post*, September 13, 2016, https://www.huffingtonpost.com/entry/mass-incarceration-cost_us_57d82d99e4b09d7a687fde21.

10: ALEXA, CALL FOR HELP

"The unexpected use of this new technology to contact emergency services..." Morgan Winsor, "Smart Home Device Alerts New Mexico Authorities to Alleged Assault," ABC News, July 6, 2017, https://abcnews.go.com/US/smart-home-device-alerts-mexico-authorities-alleged-assault/story?id=48470912.

Wired *magazine explains why*... Emily Dreyfuss, "An Amazon Echo Can't Call the Police—But Maybe It Should," *Wired*, July 16, 2017, https://www.wired.com/story/alexa-call-police-privacy/.

What is the language of reason and plain sense upon this same subject... Jeremy Bentham, "Critique of the Doctrine of Inalienable, Natural Rights" [original title "A Critical Examination of the Declaration of Rights: Preliminary Observations"], excerpt from *Anarchical Fallacies,* in *The Works of Jeremy Bentham*, vol. 2, ed. John Bowring (Edinburgh: William Tait, 1843), http://fs2.american.edu/dfagel/www/Class%20Readings/Bentham/AnarchichalFallicies_excerpt.pdf.

11: SCHOOL SAFETY

Christopher Ingraham, a reporter for the Washington Post, *studied the FBI's nationwide data*... Christopher Ingraham, "Guns in America: For Every Criminal Killed in Self-Defense, 34 Innocent People Die," *Washington Post*, June 19, 2015, https://www.washingtonpost.com/news/wonk/wp/2015/06/19/guns-in-america-for-every-criminal-killed-in-self-defense-34-innocent-people-die/.

... studies, such as one conducted by the National Academies' Institute of Medicine and National Research Council... National Academies Institute of Medicine and National Research Council, *Priorities for Research to Reduce the Threat of Firearm-Related Violence* (Washington, DC: National Academies Press, 2013), https://www.nap.edu/read/18319/chapter/3.

Sheldon Greenberg, a Johns Hopkins professor and a former police officer... Mimi Kirk, "What Research Says about Arming Teachers," CityLab, March 14, 2018, https://www.citylab.com/life/2018/03/what-the-research-says-about-arming-teachers/555545/.

A 2018 report from CNN... Chip Grabow and Lisa Rose, "The US Has Had 57 Times as Many School Shootings as the Other Major Industrialized Nations Combined," CNN, May 21, 2018, https://www.cnn.com/2018/05/21/us/school-shooting-us-versus-world-trnd/index.html.

Take the example of Australia... Matthew Grimson, "Port Arthur Massacre: The Shooting Spree That Changed Australia's Gun Laws," NBC News, July 25, 2015, https://www.nbcnews.com/news/world/port-arthur-massacre-shooting-spree-changed-australia-gun-laws-n396476.

That's 88.8 guns per 100 people... Esther Zuckerman, "There Are 88.8 Guns per 100 People in This Country: That's the Highest Gun-Ownership Rate in the World," *Atlantic*, December 14, 2012, https://www.theatlantic.com/national/archive/2012/12/guns-per-person-usa/320459/.

Since 1968, there have been an estimated over 1.5 million deaths from firearms... Louis Jacobson, "More Americans Killed by Guns since 1968 than in All U.S. Wars, Columnist Nicholas Kristof Writes," Politifact, August 27, 2015, https://www.politifact.com/punditfact/statements/2015/aug/27/nicholas-kristof/more-americans-killed-guns-1968-all-wars-says-colu/.

Omar Mateen, who killed 49 people and wounded 58 others... Andrew Gumbel, "Mass Shootings: Why Do Authorities Keep Missing the Warning Signs?," *Guardian,* March 6, 2018, https://www.theguardian.com/us-news/2018/mar/06/mass-shootings-fbi-law-enforcement-prevention.

DARPA *Grand Challenge...* Defense Advanced Research Projects Agency (DARPA), "The Grand Challenge," https://www.darpa.mil/about-us/timeline/-grand-challenge-for-autonomous-vehicles.

12:CALLING FOR PROGRESSIVE ACTIVISTS

*progressive: /pr*uh-**gres**-*iv/...* Definition from https://www.dictionary.com/browse/progressive.

Several leading human rights organizations led an intensive, multi-year campaign to ban the use of the TASER *technology...* That also ignored the fact that, if you were looking to torture someone, a TASER device would be a very expensive, ineffective choice. For those who don't know, a TASER weapon typically costs around one thousand dollars. A good part of that expense is to fund the medical safety and other research that goes into ensuring a safe design. Its primary advantage is that a TASER weapon can fire projectiles and connect to a resistant or moving person, at a cost of about thirty-five dollars per shot. The electric signals sent by a TASER are designed to incapacitate someone out of your control, briefly. By definition, if you are torturing someone, you have them under control already, and your intention is normally to cause them pain and discomfort for as long as is necessary. Importantly, the TASER system has one other drawback for would-be torturers: it keeps a usage log of every time the trigger is pulled. This is not a helpful feature if you want to abuse someone with it and later deny it to a tribunal in The Hague.

... since 2000, there have been more than a hundred officer-involved shootings in San Francisco... This data was drawn from two datasets available at San Francisco Police Department, "White House Police Data Initiative," https://sanfranciscopolice.org/data.

13: OUR BIGGEST PROBLEMS REQUIRE OUR BRIGHTEST MINDS

... if the smartest people choose not to work on police, military, and defense technology, that necessarily means that those problems will be confined to a less-smart crowd... To be clear, there are many smart people working at government agencies and companies that support the Department of Defense. This is not to say that all of the smartest people work in Silicon Valley, but that a large cohort of the most technically adept people work in these companies and are systematically removed from working on the problems of public safety, military, and violence.

"America, we have a problem," said Representative Jackie Speier... Scott Shane, "These Are the Ads Russia Bought on Facebook in 2016," *New York Times,* November 1, 2017, https://www.nytimes.com/2017/11/01/us/politics/russia-2016-election-facebook.html.

"If big tech companies are going to turn their back on [the] US Department of Defense..." Levi Sumagaysay, "Jeff Bezos: Tech Companies Shouldn't 'Turn Their Back' on U.S. Government," *Mercury News*, October 16, 2018, https://www.mercurynews.com/2018/10/16/jeff-bezos-tech-companies-shouldnt-turn-their-back-on-u-s-government/.

14: GOVERNMENT IN A WORLD WITHOUT PRIVACY

As reported in an April 2018 interview in Wired *magazine...* Steven Levy, "Cracking the Crypto War," *Wired*, April 25, 2018, https://www.wired.com/story/crypto-war-clear-encryption/.

Beginning in 2014, Tim Clemans began sending release requests... Bill Schrier, "The Future of Police Video: Inside the Seattle PD's Workshop on Wearable Cameras," *GeekWire*, June 23, 2015, https://www.geekwire.com/2015/the-future-of-police-video-inside-the-seattle-pds-workshop-on-wearable-cameras/.

15: ENGINE OF VIOLENCE: ENDING THE WAR ON DRUGS

... the largest prison population on Earth... Michelle Ye Hee Lee, "Yes, U.S. Locks People Up at a Higher Rate Than Any Other Country," *Washington Post*, July 7, 2015, https://www.washingtonpost.com/news/fact-checker/wp/2015/07/07/yes-u-s-locks-people up-at-a-higher-rate-than-any-other-country/.

"As an investment, the war on drugs has failed to deliver any returns..." Charles Riley, "Richard Branson: Decriminalize Drug Use Worldwide," CNN Money, October 19, 2015, https://money.cnn.com/2015/10/19/news/richard-branson-decriminalize-drugs/index.html.

... the United States incarcerates 716 out of every 100,000 people... Lee, "Yes, U.S. Locks People Up at a Higher Rate Than Any Other Country."

A 2014 study of anti-drug programs of different countries by the Home Office in the United Kingdom... "No Link between Tough Penalties and Drug Use: Report," BBC News, October 30, 2014, https://www.bbc.com/news/uk-29824764.

In Portugal, a breakthrough model has shown... Susana Ferreira, "Portugal's Radical Drugs Policy Is Working: Why Hasn't the World Copied It?," *Guardian*, December 5, 2017, https://www.theguardian.com/news/2017/dec/05/portugals-radical-drugs-policy-is-working-why-hasnt-the-world-copied-it.

INDEX

ABOUT THE AUTHOR

ICK SMITH founded Axon Enterprise (formerly TASER International) in 1993. As the TASER device became ubiquitous in law enforcement, Smith pushed the company beyond weapons technology and toward a broader purpose of using hardware, software, and artificial intelligence to make the world a safer place. Under his leadership, Axon has grown from a garage in Tucson to a NASDAQ-listed global market leader in conducted electric weapons, body-worn cameras, and software. Smith graduated from Harvard University with a BA in biology and later earned a master's in international finance from the University of Leuven in Belgium and an MBA from the University of Chicago. Learn more at www.EndOfKilling.com.